Júlia Soares Pereira
Viviane Farias Silva
Maria S. S. Farias

Environmental Impacts in Preservation Areas in the Semi-Arid Region of Paraiba

Júlia Soares Pereira
Viviane Farias Silva
Maria S. S. Farias

Environmental Impacts in Preservation Areas in the Semi-Arid Region of Paraiba

ScienciaScripts

Imprint
Any brand names and product names mentioned in this book are subject to trademark, brand or patent protection and are trademarks or registered trademarks of their respective holders. The use of brand names, product names, common names, trade names, product descriptions etc. even without a particular marking in this work is in no way to be construed to mean that such names may be regarded as unrestricted in respect of trademark and brand protection legislation and could thus be used by anyone.

Cover image: www.ingimage.com

This book is a translation from the original published under ISBN 978-3-330-76177-3.

Publisher:
Sciencia Scripts
is a trademark of
Dodo Books Indian Ocean Ltd. and OmniScriptum S.R.L publishing group

120 High Road, East Finchley, London, N2 9ED, United Kingdom
Str. Armeneasca 28/1, office 1, Chisinau MD-2012, Republic of Moldova, Europe
Managing Directors: Ieva Konstantinova, Victoria Ursu
info@omniscriptum.com

Printed at: see last page
ISBN: 978-620-8-37998-8

Authors: Júlia Soares Pereira[1]

Viviane Farias Silva[2]

Maria Sallydelândia Sobral de Farias[3]

1 (2): PhD student in Agricultural Engineering at the Federal University of Campina Grande - Paraíba. E-mail: julia_eng@hotmail.com/flordeformosur@hotmail.com

(3): Post-Doctorate in Agricultural Engineering. Professor at the Agricultural Engineering Academic Unit of the Federal University of Campina Grande - Paraíba.

SUMMARY

SUMMARY

An Environmental Protection Area (APA) is an extensive natural area intended for the protection and conservation of the biotic (fauna and flora), aesthetic or cultural attributes that exist there, which are important for the quality of life of the local population and for the protection of regional ecosystems. The main objective of an APA is the conservation of natural processes and biodiversity, by guiding, developing and adapting the various human activities to the environmental characteristics of the area. As a conservation unit in the sustainable use category, the APA allows human occupation. These units exist to reconcile the orderly human occupation of the area and the sustainable use of its natural resources. The idea of sustainable development guides any and all activities carried out in the area. With this in mind, the aim of this work was to diagnose the Environmental Preservation Area (APA) of the Fomento Post, located at the Centre for Agricultural and Environmental Sciences (CCAA) of the State University of Paraíba Campus II, which is also home to the Assis Chateaubriand Agricultural School, characterised by the practical experiments carried out by technical students and undergraduates, which is located around two communities, Vila Florestal and Ipuarana. The area's degradation problems were assessed through photographic records and the quality of the water from a spring within the reserve and from a reservoir close to the community. It was found that in many parts of the area there is a high degree of degradation directly linked to the inhabitants around the reserve, through the exploitation of natural resources and the dumping of waste. The water analysed falls into class C2, water of medium salinity (EC between 250 and 750 p.S.Cm-1 at 25°C), and can be used whenever there is a degree of leaching. Plants with moderate salt tolerance can be grown in most cases without the need for special salinity control practices, showing good quality for irrigating existing crops.

Key words: environmental degradation, solid waste, water

CHAPTER 1

INTRODUCTION

Deforestation and the inadequate use and management of natural resources have caused, among other effects, the deterioration of arable soils, alterations to drainage networks and qualitative and quantitative losses of water in rivers, lakes and reservoirs. Inappropriate occupation favours erosion processes, reducing soil productivity and consequently transporting and accumulating sediment in reservoirs, reducing the quantity and quality of water (SANTOS, 2010).

Over the years, permanent preservation areas (APPs) on the banks of watercourses have suffered degradation, especially in urban areas, with the partial or total removal of vegetation in this strip, which should be kept intact because it guarantees the preservation of water resources, geological stability and biodiversity.

According to Martins (2001), the vegetation that develops in the permanent preservation strip along watercourses is commonly called riparian forest, represented by narrow strips of native vegetation. One of its functions is to hinder the silting up of riverbeds by not allowing the sediment carried by rainwater to reach them in its full age, altering the course of the riverbed. In addition, the roots of the plants serve to fix the banks and protect them from intense erosion.

According to Lima et al. (2000), forest hydrology can be understood as the area of human knowledge that is concerned with the environmental management of the micro-watershed. Within this, riparian forests occupy the most dynamic areas of the landscape, making them important for the quantity and quality of water in the micro-basin, since the destruction of riparian forest, in the medium and long term, reduces the storage capacity and, consequently, the flow of rivers in the dry season.

According to Assunção et al. (1996) awareness of the need to conserve and defend the environment for present and future generations is indisputable; knowledge of the law to understand the duties and prerogatives of citizens is essential; the importance of starting environmental education in childhood is unquestionable; making children understand environmental laws in order to contribute to the preservation and defence of the environment as responsible citizens; creatively changing the behaviour of as many students as possible and making them agents of defence for an ecologically balanced and healthy environment.

According to Martins (2001), the success of a riparian forest recovery project should be assessed using recovery indicators. Through these indicators, it is possible to define whether the project needs to undergo further interference or even be redirected, with the aim of speeding up the process of succession and restoration of the riparian forest's functions, as well as determining the moment when the planted forest becomes self-sustaining, dispensing with anthropic interventions.

In this context, this research is in line with the recommendations of projects to recover degraded areas, where the main way of guaranteeing the long-term sustainability of restoration models is continuous monitoring, so that measures can be indicated that will contribute to remedial actions that feed back into the project.

CHAPTER 2

LITERATURE REVIEW

2.1. Environmental preservation areas

The environment can be conceptualised, according to Art. 1998, as an agglomeration of circumstances in which all living beings interact, thus the environment is defined as the union of all external components such as climate, soil, water and other natural resources existing in the area, providing conditions for the existence of living beings. Some authors such as Santos (1996) describe that the environment involves sustaining life, be it physical or material, making it possible to remain or survive in a given place.

According to Gliessman (2000), the location or environment of a species can be based on the set of external and internal, biotic and abiotic characteristics that influence its development, so the environment must be active, i.e. in permanent movement of environmental modifications and relationships, being a very heterogeneous and important system for nature. As also reported by Monteiro (2002), the environment incorporates all living and non-living beings, which are parts that influence the ecosystem.

The environment, or scientifically defined environment, is based on concepts relating to biological, geographical and physical characteristics interacting with species, thus the formulation of the environment has a strong influence from the field of environmental sciences, which tends towards biotic and abiotic environments as stated (CALDART, 2012).

Dulley (2004) states that each living being exists in different environmental conditions depending on the relationship between the relevant environmental factors to stay alive and to fulfil their primordial metabolic needs.

Sites encompassing their natural resources and characteristics are considered Conservation Units (CU) with the aim of maintaining biological heritage. These CUs enable the sustainable use of natural resources and the sustainable economic development of surrounding populations, and can be legally created by the federal, state or municipal government (MMA, 2016).

Conservation units can be divided into integral protection units and sustainable use units, with the environmental preservation area incorporated as a category in the sustainable use unit, according to MMA (2016).

According to Esteves (2011), the environmental preservation area or environmental protection area (APA) can be conceptualised as an area of the Federal Conservation Unit for sustainable use in human occupation. These APAs are set up with the aim of preserving biological diversity and regularising occupation in order to guarantee environmental sustainability.

The environmental protection area (APA) can be defined, according to the MMA (2016) as an area with natural, aesthetic and cultural characteristics that are considerable for the quality of life of the population, it is usually a wide area, with the purpose of ensuring biological variety, and can be public or private land, while according to Louzada et al. (2009) permanent preservation areas (APP) are relevant for the conservation of natural resources, these areas being reduced due to anthropic actions that contribute to environmental degradation (OLIVEIRA et al., 2007).

According to Law 1876/99, art.3, II, a Permanent Preservation Area is a protected area, which may or may not be covered by native vegetation, with the aim of maintaining natural resources, such as water resources, the landscape, biodiversity, fauna and flora, certifying favourable human conditions.

Permanent preservation areas are places where no activities of any kind can take place, and the use of their natural resources is forbidden, even in a sustainable way. According to Catelani & Batista (2007), these areas were created to protect natural resources, i.e. the environment, by preventing human interference in their use.

The attribution of APPs is very broad, due to the importance of these areas in the conservation of populations and natural resources, not limited to the preservation of biodiversity, guaranteeing quality of life (MMA, 2011). With the aim of providing a stable environment for common use, Borges et al. (2011) assures that permanent preservation areas are continuously connected to environmental attributions.

According to Ab'Saber (2001), riparian forests have an active role in relation to the alluvial or flood plain, i.e. the river's largest reader and the drying bed, defining riparian forest as any type of tree vegetation on the banks of watercourses.

In order to analyse the environment, it is necessary to have information on how to use the techniques that must be observed, such as form, function, structure and others. Thus Verdum and Fontoura (2009) conceptualise form as a visual characteristic that must be known in the field through photographic records and geoprocessing software, while function must be understood by activities, structure is a criterion that must be linked to function, with values and functions in their historical diversity and the dynamic criterion is an activity that is always continuous and the results of this dynamic continuous flow is what must be observed.

Studying the legislation on Permanent Preservation Areas, Sauer (2008) states that under the law these areas in front of environmental activities are only considered PPAs when declared by the government, without technical criteria to specify the areas.

According to article 271965 of the Forest Code, environmental preservation areas (APP's) are considered to be forests and the following natural vegetation that is located in the courses and contours of any watercourse, springs, hilltops, mountains and hills, slopes with a gradient greater than 45° and in fields, native forests and vegetation located at an altitude of more than 1,8000 metres, as described in Brazil (1965).

This definition given by Federal Law No. 4.771/1965 continues in the current Federal Law No. 12.651/2012 and amendments, considered the New Forest Code, and the Legal Reserve was also established by legislation, with the aim of providing for the preservation of biodiversity and natural resources on rural properties.

The law (Federal Law No. 12,651/2012) defines a legal reserve as an area located inside a rural property or possession with the function of ensuring the sustainable economic use of natural resources, providing for the conservation and rehabilitation of ecological processes, promoting biodiversity conservation, sheltering and protecting wild fauna and native flora (Article 3, III).

Some authors (MILANO, 1992; DETZEL, 1992; SATTLER 1992;

CAVALHEIRO, 1992) state that there is a favourable intervention of the vegetation cover associated with the dynamic activity of the urban environment, highlighting the climatic factors, due to the reduction of temperatures, air pollution and acoustics, aesthetics, psychological effects, environmental comfort, among other qualities that improve the quality of life in urban areas. Bernatzky (1982) reported that a barrier with approximately 30 metres of vegetation in the middle of an industrial site and a residential area, the particulate matter coming from the industries was captured 100% by the vegetation and reduced the gaseous pollutants, noting the importance of vegetation for the population.

So sustainable development can be achieved when there is awareness that natural resources may come to an end, i.e. they are finite, and the need for their use must be managed correctly, as Nascimento (2012) states.

2.2. Environmental degradation and sustainability

Environmental degradation is considered to be the negative alteration of environmental attributes, as stated in the National Environmental Policy, according to Law No. 6.938/81, art. 3, item II. Meneguzzo (2006) conceptualises the origin of environmental degradation and Sanchez (2008) further complements this rationale of environmental degradation as peculiarities of unfavourable environmental impacts.

Lemos (2001) describes environmental degradation as the devastation or wear and tear caused to the natural environment as a result of socio-economic, population and biological activities. Authors such as Leite et al. (2011) base the definition of environmental degradation on another view, stating that it can be produced through agricultural activity due to the enormous changes that are necessary for this activity, such as changes in land use, soil compaction, production of waste polluting the area, rivers and soils, deforestation and desertification, in other words, it is a practice that largely degrades the environment.

Degraded systems are conceptualised as insufficient due to their lack of conditions for self-recovery, which would make it possible to make them sustainable by inserting them into built, cultivated or modified systems (XAVIER, 2010).

Pinto and Coronel (2014) report that in Brazil, despite the negative impacts that occur when carrying out this activity, farming also contributes to job creation, food supply and other positive factors, so it is important to interact with sustainable alternatives for this degradation-providing activity, such as the appropriate final destination for the waste generated, planning the cattle grazing area and rotating the areas, as well as other solutions that can be applied to this activity specifically.

Silva et al. (2012) state that the unlimited exploitation of the world's natural resources is the primary factor in the occurrence of environmental degradation, reporting that soil erosion is an element that is directly related to the chain of environmental degradation. This indiscriminate exploitation occurs on a larger scale than the ability of natural resources to regenerate and, as a consequence, these resources become scarce, as Rossato (2006) states. Thus, environmental indicators are studied in order to observe the interaction of these ecosystems and their activities with man's use of natural resources (LIRA and CÂNDIDO, 2008).

The National System of Conservation Units, established by Law No. 9.985/2000, is based on the concept of nature conservation, which encompasses the form of management that human beings apply to nature, addressing preservation and sustainability, making it possible to maintain, restore and recover the environment, providing availability for the current and future population, ensuring the survival of all living beings.

When an area is suffering from very aggressive environmental impacts that compromise its natural recovery capacity, this is a more dangerous form of environmental degradation, as explained by Sampaio et al. (2005). Desertification can thus be the result of environmental degradation, making it important to carry out studies on environmental factors, their interaction and perceptions of sensitivity in relation to the impacts caused to the environment.

Desertification can be defined according to Saadi (2000) as a chain of factors that cause deterioration of environmental conditions to an advanced stage that disrupts the basic requirements of life for the population, thus the use of inadequate irrigation can cause salinisation, generating desertification (SOUSA, 1985). As for the causes that influence the occurrence of desertification, Carvalho and Almeida (2007) report that the practice of activities such as ploughing and burning accentuates the desertification process. These practices have been carried out by man for several decades until the present day, and are the main motivation behind desertification, as mentioned by Muniz (2004).

Research into environmental degradation is based on indices carried out throughout Brazil (PAIS et al., 2012; Lemos, 2001). The reduction in vegetation cover exposes the soil to the impacts of rain droplets and winds, causing soil erosion, reduced infiltration and increased surface runoff, thus increasing the loss of soil and nutrients through erosion, as Gutierres and Squeo (2004) state that soils without vegetation cover are more prone to erosion, with significant amounts of soil being lost. Erosion can occur in a variety of situations, whether natural or man-made, and is classified as water or wind erosion. Wind erosion is when the wind passes through and carries the surface grains from one place to another, according to Vieira and Lima (2007). Soil erosion affects the quality of life of the population, due to the reduction of nutrients and loss of plant species with deterioration of water resources, which are environmental damages (SUERTEGARY, 1996).

As we have been contextualising the different forms of environmental degradation, those caused by human action are the most aggressive towards natural resources, making agriculture an enemy of nature and one of the most obvious causes of wear and tear on the soil and natural resources, as Silva and Corrêa (2007) explain, the unsustainable way of agricultural production with physical systems generates various types of problems for the environment that are visually apparent.

Indiscriminate exploitation of natural resources, especially the soil, modifies the environment when practised in an abusive way, so sustainable forms become an alternative in the quest to reduce environmental impacts. Firmino and Fonseca (2008) warn that agricultural practices directly affect natural resources. As Lima et al. (2004) also state when they describe that when farmers and small breeders do not apply appropriate methods in their activities, they exceed the use of resources, contributing to increasing desertification through the great environmental degradation that has occurred.

The incorrect use of natural resources, according to Barros and Silva (2010), is one of the main causes of environmental degradation, causing instability in the constituents, both biotic and abiotic, and this topic is widely significant in the debate on environmental issues aimed at sustainability.

Capitalism is a way of using all natural resources in the pursuit of economic development, which leads to the application of other related terms such as globalisation. In this way, with increased exploitation of the environment in order to raise economic indexes, there is growing social inequality, favouring environmental degradation in poorer areas (ASSIS, 2003).

Agricultural production in Brazil, due to the method of management applied, is an aggressive environment that causes environmental damage, such as water and soil pollution, a reduction in available natural resources, as well as generating waste and releasing it into the environment without any prior treatment (KOZIOSKI AND CIOCCA, 2000).

In the search for sustainable production systems, with lower rates of degradation and greater recycling of the waste produced, there is a greater global focus on environmental situations in the political, social and economic spheres, as discussed by Capper et al. (2009). Due to so many concerns about the environmental degradation caused by man's activities, researchers have used their studies to show results on the value of agricultural production per area, among other environmental factors, such as plant diversity, biome preservation, which in general terms assesses the qualitative rather than the quantitative magnitude of the damage done to the environment, as analysed by Araujo Filho et al. (2006).

Moraes and Jordão (2002) found that the degradation of water resources affects human health. These same authors state that water withdrawn for use returns to water bodies without any prior treatment and with poor quality, so the waste products contained in the water after use influence the degradation of the water environment. When there are changes to the natural system, these sites are considered to be negatively affected, according to Silva et al. (2011).

According to Oliveira and Souto (2011) an extensive area is degraded through inadequate management by man. Machado et al. (2013) reports that environmental regulations in Brazil underwent a number of changes at the end of the 20th and 21st centuries, but there is still a lack of regulations dealing specifically with soil pollution. Avanzi et al. (2009) found that the Brazilian Forestry Code aimed to maintain the soil's vegetation cover, thus conserving water and preventing erosion and degradation.

There are different definitions of sustainable development. Goodland and Ledoc (1987) interpret the term sustainable development as a model of socio-economic modifications to improve the utilisation of available resources in order to maintain their potential in the future. These same authors say that sustainable development was initially intended to achieve an economic balance that does not deplete existing natural resources and that this economic model should be continuously disseminated throughout all generations to enable favourable living conditions for all living beings, using renewable resources without degrading, reducing or eliminating them from nature. Allen (1980) also states that sustainable development means the supply and quality of human needs in order to survive.

Characterised by proportionality throughout its process, according to Canepa (2007) sustainable development has to maintain a balance in all its phases in order to provide future generations with a minimum of natural resources to continue living a quality life.

Barbosa (2008) states that the term sustainable development appeared in research carried out by the United Nations Organisation (UNO) as a way of providing information to the population, which was experiencing a socio-environmental crisis in the 20th century.

Marzall and Almeida (2000) explain that lately, around the second half of the last decade in particular, researchers and interested parties have been looking into sustainability indicators to help them understand the best methods to take, so many meetings have been held to discuss the parameters to be studied among researchers in the field. However, there is no in-depth breakdown of total indicators in the scientific community, but studies are geared towards soon having a list available of positive and/or negative indicators of sustainability, whether environmental, economic or social.

Environmental criteria are a way of acquiring real data in relation to a given location, as Mitchell (1996) explains. While Mueller et al. (1997) defines an indicator as individual or collective information, for an indicator to be considered good it must be easy to understand, quantify, analyse and interact with the state of the phenomenon being verified. It is a parameter that shows the condition of the system (SHIELDS et al. 2002). Dumanski (1994) described few indicators that are simple to understand and apply, as they are based on identifying natural factors and are used by farmers.

Studying the challenge of sustainable development, Barbosa (2008) concluded that this term, due to its importance, has been used as a political promise, but that attitudes must be practised to alleviate the degraded environmental conditions resulting from the consumerism of today's humanity. This author also describes sustainability as a way of trying to provide a symmetrical distribution of resources that is viable.

Nascimento (2012) reflects on the dimensions of sustainability and explains that environmentally sustainable development is a process of equivalent production and consumption, i.e. producing and consuming in such a way as to ensure that ecosystems are able to recover themselves. This is the first dimension mentioned by the author. The second dimension considered is the economic one, which envisages an increase in income and the preservation of environmental resources, especially non-renewable ones such as fossils. The last dimension mentioned is the social one, which imagines a sustainable society where people are environmentally aware, with the minimum necessary for a quality life.

Reflection on social practices, in a context marked by the permanent degradation of the environment and its ecosystem, involves a necessary articulation with the production of meanings about environmental education. The environmental dimension is increasingly becoming an issue that involves a range of actors in the educational universe, making it possible to engage the various knowledge systems, training professionals and the university community from an interdisciplinary perspective. In this sense, the production of knowledge must necessarily take into account the interrelationships between the natural and social environments, including an analysis of the determinants of the process, the role of the various actors involved and the forms of social organisation that increase the power of alternative actions for a new development, from a perspective that prioritises a new development profile, with an emphasis on socio-environmental sustainability (JACOB, 2003).

According to Campos (2008), environmental education is the permanent educational action through which the community becomes aware of its global reality, the type of relationships that people establish between themselves and with nature, the problems arising from these relationships and their root causes. It develops, through a practice that links the student with the community, values and attitudes that promote behaviour aimed at overcoming the transformation of this reality, both in its natural and social aspects, developing in the student the skills and attitudes necessary for this transformation. These are the processes through which individuals and the community build social values, knowledge, skills, attitudes and competences aimed at conserving the environment, which is a good for the common use of the people, essential to a healthy quality of life and its sustainability. It is a comprehensive form of education that aims to reach all citizens, through a permanent participatory pedagogical process that seeks to instil in the student a critical awareness of environmental issues, with critical awareness being understood as the ability to grasp the genesis and evolution of environmental problems.

There is therefore a need to increase the means of information and access to them, as well as the inductive role of public authorities in educational content, as possible ways to change the current situation of socio-environmental degradation. It is a question of promoting the growth of environmental awareness, expanding the possibility for the population to participate at a higher level in the decision-making process, as a way of strengthening their co-responsibility in monitoring and controlling the agents of environmental degradation (JACOB, 2003).

2.3. Water Resources in Semi-Arid Regions

Drought is a natural disaster considered by many to be the most complex, yet the least understood of natural hazards and the one that affects more people than any other hazard (HAGMAN, 1984). Droughts have causes of planetary proportions and are influenced by various factors, among which it is worth highlighting: the difference in surface temperature of the waters of the North Atlantic, which are warmer, and those of the South Atlantic, which are colder; the shift of the Intertropical Convergence Zone to the Northern Hemisphere at times when it is expected to remain in the South; and the appearance of the phenomenon known as El Nino, characterised by a rise in temperature in the Eastern Equatorial Pacific Ocean. In addition to all this, the shape of the relief in the Northeast and the high reflectivity of the crust are the main local factors inhibiting rainfall production (SILVA, 2003).

A large part of the earth's surface is covered by areas of arid and semi-arid climates, corresponding to a third of the globe's emersed land, according to Conti (1998), equivalent to 50,000,000 km^2 . Cited by Melo Filho (2006), Raya says that the world's semi-arid surface can be estimated at between 10 and 13 per cent of the planet's land.

These regions are home to more than 1 billion inhabitants or more than 20% of the world's population, who live in environments susceptible to anthropogenic actions, where agricultural investment is one of the most costly in the world, thus posing serious problems for the local population, given that the inhabitants of these regions rely on primary activities as their main source of economic support. Although most of our planet's surface is made up of water (three quarters), 97% of this is concentrated in seas and oceans, leaving only 3% of fresh water, 75% of which forms the polar ice caps, and of the remaining 25%, 98.8% is groundwater, leaving only 1.2% of surface water (rivers and lakes) easily accessible for human consumption (CEBALLOS, 1995).

In the north-eastern region of Brazil, water is one of the limiting elements to its development. The rainfall that occurs there is irregular both temporally and spatially. The high rate of evaporation and the crystalline geological nature of most of its terrain, combined with the growing demand for water for human and animal supply and irrigation, among others, give this region a negative annual water balance (SALATI et al., 1995).

The Brazilian semi-arid region is one of the largest, most populous and wettest in the world. It extends over 980,089.26 km^2 and covers 1,135 municipalities in the states of Piauí, Ceará, Rio Grande do Norte, Paraíba, Pernambuco, Alagoas, Sergipe, Bahia, Minas Gerais, Espírito Santo and Maranhão. This region is home to 21,718,168 people (CARVALHO, 2006).

The climate is remarkable, with an average rainfall of 500 mm/year in two distinct seasons, one with rainfall of 3 to 5 months in the first half of the year called winter, and another dry season that lasts 7 to 9 months and can be extended, called summer. Due to these climatic characteristics, the Northeast region has one of the highest evaporation rates in Brazil, because of its low latitude, it is very close to the equator, located between 1 °S and 18 °S, where the sun's rays fall almost vertically from January to December, providing high thermal averages and luminosity (2,800 hours) throughout the year. The winds are dry, hot and have a high average speed (15 to 25 km/h), and the temperature reaches an average of 23 to 27°C, considered very high, a condition that favours high evaporation and reaches figures of around 2,000 mm/year.

In the semi-arid region, water policy has prioritised the construction of hydraulic works, such as dams of varying sizes, which, because they are located on private property, restrict their use by the entire population, while water reservoirs with greater storage potential are managed by the state to guarantee their use by the population, as Garjulli (2003) reports.

The soils are generally shallow and stony, poor in organic matter, with a low capacity to accumulate water, although they are rich in soluble mineral salts, especially calcium and potassium, and have a neutral or near-neutral pH.

The semi-arid northeast of Brazil is one of the hottest dry regions on the planet (MENDES, 2006). However, the concept of soil, livestock and people being punished by drought is changing. With the development of irrigation, the use of drought-tolerant species and livestock production, the desolate panorama is giving way to a production scenario where available technologies have made it possible to commercially produce various agricultural and livestock products capable of competing on national and international markets.

According to Lucena (1998), around 60% of the Northeast Region corresponds to the area where droughts occur and is known as the Drought Polygon, but in this polygon there are regions whose rainfall exceeds 800 millimetres, a type of high-altitude microclimate called "Brejo", with abundant rainfall, milder temperatures, higher humidity, frequent cloud cover and cerrações. For Cadier (1994), the climate of the brejos cannot be described as semi-arid, as it plays a major role in the regional economy and originates or sustains the sources of many rivers.

The semi-arid region of Paraíba covers an area of approximately 20,000 km^2 and is characterised, from a geo-environmental point of view, by the diversity of its landscapes, with the semiarid conditions that affect a large part of its territory and the high spatial and temporal rainfall variability inherent to this type of climate being a striking feature of the region's natural framework (SALES, 2002).

Vicente and Carneiro (1993) cite that approximately 99% of the territory of Paraíba is affected by periodic droughts, which covers an area of 56,439.838 km2 and has an estimated population of 3,623,215 inhabitants in 2006 (IBGE, 2006).

According to the GOVERNMENT OF THE STATE OF PARAÍBA (1985), the hydrography is characterised by the predominance of temporary rivers, with flow regimes dependent on the amount and distribution of rainfall.

2.4. Floristic Resources of the Semi-Arid

The Caatinga biome is made up of dry, deciduous and thorny forest vegetation that covers most of the Brazilian semi-arid region (SAB), almost all of which is in the north-east, as it includes part of the north of Minas Gerais, occupying an area of approximately 1.08 million km^2 , with a population of 24.6 million inhabitants, making it the most populous semi-arid region in the world (EMBRAPA, 2000). In the indigenous language, Caatinga means "white forest" because, during the dry season, most of the plants lose their leaves, giving the whole landscape a white appearance due to the white or silvery colour of the bark (MAIA, 2004).

The association of edaphic factors and anthropogenic actions produce mosaics of vegetation, causing the caatinga to show enormous spatial and temporal variability, even when seen through the stereotype of a rock outcrop where xique-xique and macambira predominate (MIRANDA, 1986). Araújo Filho (1986) explains that this variability in the composition and arrangement of its botanical components is due to responses to succession processes and various environmental factors, where plant density, floristic composition and the potential of the herbaceous layer vary depending on the characteristics of the soil, rainfall, altitude, relief, anthropogenic actions, etc.

Environmental degradation in the northeastern semi-arid region, although it has occurred slowly over 300 years of inadequate exploitation, is often associated with the combination of very low rainfall that varies greatly in time and space, and evapotranspiration that is often very high. This has also led to a high risk of agricultural activity at a very low technological level (ARAÚJO et al., 1995).

Queiroz et al. (1993) identified four causes of the disappearance of native vegetation in the Brazilian Semi-Arid Tropics in the most recent period: 1) the formation of pastures; 2) the implementation of irrigation projects; 3) energy production for various activities, such as potteries, bakeries, plaster and lime production, and 4) natural or human-induced fires. These causes, together or in isolation, have contributed not only to reducing the secular utilisation of some fruit and/or fodder and/or timber species, but also to the disappearance of the genetic variability of some and the near extinction of others.

The systematic elimination of vegetation cover and the improper use of land have led to serious environmental problems in the semi-arid northeast, including a reduction in biodiversity, soil degradation, compromised production systems and the desertification of large areas in most of the states that make up the region (BRASIL, 1995).

In the northeastern semi-arid region, the degradation of natural resources and especially the decrease in soil fertility have been caused by the increased intensity of land use and the reduction in native vegetation cover (MENEZES & SAMPAIO, 2002). Inadequate soil management in the semi-arid region has become one of the region's main problems and can lead to degradation of its physical, chemical and biological attributes, which are essential for maintaining vegetation (SALCEDO,

2004).

2.5. Riparian forests

For Lima and Zakia (2000), forest hydrology can be understood as the area of human knowledge that is concerned with the environmental management of the micro-watershed. Within this, riparian forests occupy the most dynamic areas of the landscape, making them important for the quantity and quality of water in the micro-basin, since the destruction of riparian forest, in the medium and long term, reduces the storage capacity and, consequently, the flow of rivers in the dry season.

According to Carpanezzi (2000), the hydrological role of the forest is the result of a network of interactions in which vegetation in general, especially the forest, allows for the infiltration and temporary storage of water in the soil, thereby controlling erosion, conserving the soil and regulating the flow of rivers.

According to Rocha (2005), spring protection crowns are defined as the area between the watershed and the base of the ravines, approximately halfway up the slope. This area is of great importance for the upwelling of groundwater, as its main characteristic is that it favours the infiltration of rainwater and also prevents erosion.

Riparian forest is a permanent preservation area, which according to the Forest Code (Law No. 4,771/1965) must be left untouched, and if it is degraded, immediate restoration must be planned. This law has existed for 40 years, but has not always been complied with. All natural vegetation (arboreal or not) along riverbanks and around springs and reservoirs must be preserved. According to Article 2 of this law, the width of the strip of riparian forest to be preserved is related to the width of the watercourse (CAMPOS, 2008).

Figure 1 shows the dimensions of riparian forest strips in relation to the width of rivers, lakes, reservoirs and springs, according to the Forest Code.

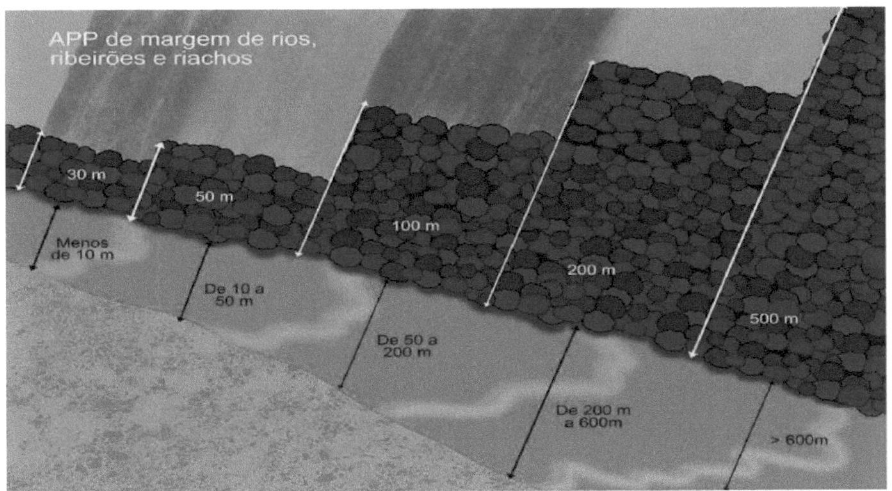

Source: PRAD de SANTOS, R. T., 2010.

Figure 1: Width of riparian forest related to springs and the width of the water mirror of rivers, lakes and reservoirs.

In general, it is recognised that riparian forests contain a set of characteristics that are unique to them. In this sense, riparian vegetation occurs on portions of land that include both the banks of a watercourse and the floodplain (LIMA, 1989). This type of vegetation is well defined in regions dominated by savannah or grassland formations, but less differentiated in regions dominated by forests, where it is mainly distinguished by its floristic composition (MANTOVANI et al., 1989).

According to Lima et al., a spring is the outcropping of the water table, which will give rise to a source of accumulated water (dam), or watercourses (streams, creeks and rivers). Because of its inestimable value within an agricultural property, it must be treated with special care. Watershed management must take into account the preservation and improvement of water in terms of quantity and quality, as well as its interferences in a geomorphological unit of the landscape, as the most appropriate form of systemic manipulation of a region's resources.

According to Rodrigues et al, springs, watercourses and reservoirs, although different from each other in terms of preservation strategies, have the basic common points of controlling soil erosion through physical structures and plant barriers, minimising chemical and biological contamination and mitigating water loss through evaporation and consumption by plants.

CHAPTER 3

MATERIAL AND METHODS

3.1. Research Material and Methods 1.

3.1.1. Characterisation of the Area

The Environmental Preservation Area shown in this research, Posto de Fomento - Produção de Mudas, as shown in Figure 2, is located at Sítio Imbaúba, S/N°, Zona Rural in the municipality of Lagoa Seca (PB) with Latitude 07°10'15", Longitude 35° 51'13" and an altitude of 634 metres, covering a total of 68.3 km^2 . Its climate is tropical humid, with an average annual temperature of around 22°C, with a minimum of 18°C and a maximum of 33°C.

Source: Google, 2017.

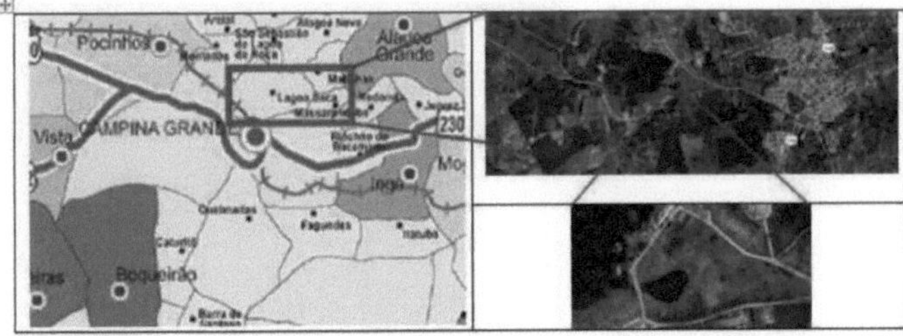

Source: Google, author

Figure 2 - Location of the environmental preservation area, Lagoa Seca/PB.

In the municipality of Lagoa Seca/PB, there is a remnant of Atlantic Forest, known as the IBAMA Forest, with continuous negative movement from human actions in an area of approximately 23 hectares, in which its floristic heterogeneity is abundant in a variety of native and exotic species, as stated by Lira (2013).

Lagoa Seca was recognised as a municipality in the state of Paraíba by state law No. 3133 of 1964 in 1958/1959 (IBGE, 2017), with a total population of 25,900 in 2010 and an estimated population of 27,398 in 2016. In 2010, the Human Development Index (HDI) was 0.627, according to the UNDP Human Development Atlas (2013). According to the 2010 Demographic Census (IBGE, 2010), there were 279 permanent private households with no bathroom or toilet, 1,870 households with rudimentary sewage, 1,541 households with septic tanks, 1,769 households with a general sewage or rainwater system, 10 households with sewage exclusively in rivers, lakes or the sea, 282 households in ditches and 1,374 households with toilets.

3.1.2. Methodological Procedures

On-site visits were made at two different times in 2010 and after six years in 2016, and through photographic records and water samples collected for analysis for irrigation purposes at the same time, the impacts on the environmental preservation area were interpreted.

The water samples used to analyse the water for irrigation purposes were collected both from one of the reservoirs near the community and within the environmental reserve itself, as shown in Figure 3.

Source: Author

3.1.3. 3. Locations (Lagoa Seca/PB) where water samples were collected for analysis in 2010.

The following procedures were followed to collect the water samples, as suggested by Almeida (2010): The water samples for collection were carried out in plastic bottles, specifically PET bottles with a capacity of one litre, which were washed three times, with the water to be sampled and in being filled completely without leaving any air. The same author highlights procedures in relation to the collection site, so when in running waters such as rivers or streams, they should be taken at a depth of an average of 12 cm above the surface and at a distance from the banks, and in reservoirs, the sample should be collected near the water intake for irrigation, procedures which were followed at the time of collection.

After 6 years, i.e. in 2016, the qualitative identification and characterisation of the environmental impacts was carried out using the Spontaneous Methodology (Ad Hoc), which is a method based on the empirical knowledge of the subject matter expert and\or the area in question. In addition, in accordance with the *aforementioned* methodology, a descriptive "check-list" was used, which is one of the most widely used methods in Environmental Impact Assessment. It consists of identifying and listing the impacts, based on a detailed environmental description carried out by specialists in the physical, biotic and socio-economic environments (Crispim et al., 2013).

Surface water was sampled by quickly immersing the collection container with the mouth down, approximately 15 cm below the water surface to avoid introducing surface contaminants. After removal, a small portion of the sample was discarded, leaving enough empty space to allow good homogenisation before starting the analysis, then the container was closed immediately and identified, according to the methodology of Vianna et al. (2011). After collecting the samples, all the material was stored in accordance with NBR 9898 (ABNT, 1987).

Two samples were taken at different points in 2016 (Figure 4) and sent to the Irrigation and Salinity Laboratory at the Federal University of Campina Grande-PB for physico-chemical analyses of the water. The assessment of water quality for irrigation followed the guidelines of Ayres and Westcot (1991) and Richards (1954).

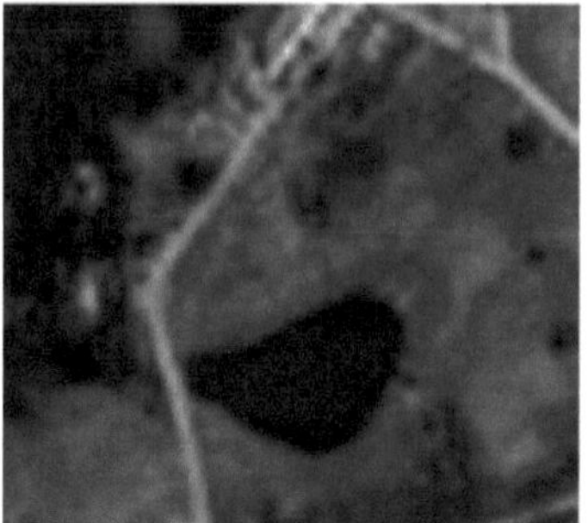

Figure 4. Geographical location of the weir in 2016.

3.2. Research Material and Methods 2.

3.2.1. Environmental characterisation of the area

The research site, the community of Uruçu, is in the municipality of Gurinhém, located in the Gurinhém Microregion and in the Agreste Paraibano/PB mesoregion (Figure 5) with an area of 309 km^2 , approximately 0.548% of the state, 0.0199% of the region and 0.0036% of the entire Brazilian national territory, with the municipal seat at an altitude of approximately 104 metres and 61.9 km from the capital. Access from João Pessoa is via the BR 230/PB 063 motorways (CPRM, 2005).

The area of the territorial unit in 2015 was 346.067 km^2 , with a population density in 2010 of 40.08 inhabitants/km^2 , according to IBGE (2016). According to the CPRM (2005), the municipality of Gurinhém is located in the Paraíba River basin, in the Lower Paraíba region.

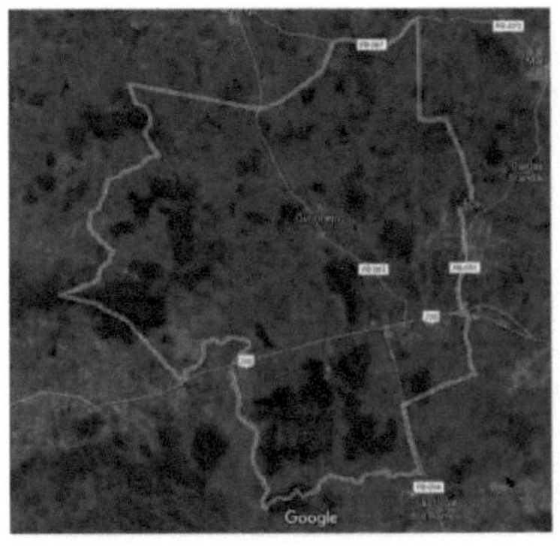

Fonte: Google, 2017

Fonte: Autor

Figure 5: Spatial location of the area surveyed, municipality of Gurinhém/PB.

The rural community of Uruçu (Figure 5) is located in the municipality of Gurinhém, approximately 80 kilometres from the city of Campina Grande/PB, accessed via the BR-230 towards the capital of Paraíba, João Pessoa. According to Oliveira et al. (2006), the Uruçu community has around 1,000 inhabitants and its main characteristic is collective organisation, with a Community Association of Rural Producers of Uruçu that has been active since 2001, made up of members who generally work on third-party properties.

Santos Junior et al. (2011) report that the productive activities carried out are focused on subsistence agriculture such as corn, fava beans, beans, sweet potatoes, as well as cotton, which is commercialised on a larger scale in the region itself. These authors also state that the area is surrounded by farms and in this location there are difficulties in obtaining land for cultivation, so it is necessary to rent for an average of one year, when farmers take possession to plant grass and graze animals.

The district of Gurinhém was created in 1873 by provincial law no. 510 and by state law in 1915. It was considered a municipality in 1958/1959 (IBGE, 2017). The total population in 2010 was 13,872, with an estimated population of 14,126 in 2016. In 2010, the Human Development Index (HDI) was 0.556, according to the Human Development Atlas - UNDP (2013). According to the 2010 Demographic Census (IBGE, 2010), there were 207 permanent private households with no bathroom or toilet, 2,766 households with rudimentary sewage, 108 households with septic tanks, 133 households with a general sewage or rainwater system, 62 households with sewage exclusively in rivers, lakes or the sea, 34 households in ditches and 508 households with toilets.

It is located in the geoenvironmental unit known as the Sertaneja Depression, with typical semi-arid vegetation made up of hyperxerophilic caatinga with a dominant gently undulating relief and a semi-arid tropical climate. The rainy season begins in November and ends in April. The average annual rainfall is 431.8 mm. With regard to the soils, in the Patamares Compridos and Baixas Vertentes of the gentle undulating relief there are Planossolos, poorly drained, with medium natural fertility and salt problems; Topos and Altas Vertentes, Brunos não Cálcicos soils, shallow and with high natural fertility; Topos and Altas Vertentes of the undulating relief there are Podzólicos, drained and with medium natural fertility and the Elevações Residuais with Litholic soils, shallow, stony and with medium natural fertility (CPRN, 2005).

3.2. 2. Methodological procedures

A questionnaire was drawn up with socio-economic questions and applied in the community. 16 residents were interviewed, including teachers and students from the Anália Arruda School, which is located in the community itself.

Three years after the first planting of the seedlings, visual indicators were assessed through current photographic records compared with records from before the project was implemented, in order to ascertain the recovery/degradation of the study area.

CHAPTER 4

MATERIAL AND METHODS

4. RESULTS AND DISCUSSION

In Lagoa Seca/PB, in the environmental preservation area, there was noticeable degradation due to the deposit of solid waste on the site (2013), as well as wear and tear of the soil due to the process of laminar erosion, caused by the exposure of the soil to the weather, whether physical, chemical or biological, directly influencing the loss of soil particles and nutrients, especially after precipitation. The area is protected by fences which, at the time of the research, could be seen to be fragile. They were made of wood and were sometimes badly damaged by weather or human activity, favouring the entry of animals and people who amplify the damage caused to the site, as can be seen in Figure 6.

In the process of reclaiming areas, one of the alternatives to be applied is fencing, which reduces negative environmental impacts. According to Resolution 429/2011, the recovery of areas can be carried out by protecting the native species of the site through the isolation or fencing of the area to be recovered, adopting the methods required for each situation (CONAMA, 2011).

In the vicinity of the environmental preservation area, there is a population that lives there and helps to aggravate the degradation that has occurred there by throwing their rubbish and solid waste directly onto the site without any prior treatment, despite the fact that there is rubbish collection in the city on exact and scheduled days that the community is informed about (FIGURE 6).

figure 6: photographic records taken in the environmental preservation area-lagoa seca/pb, 2010.

It is important to carry out work in the community, highlighting education in the countryside. In this way, environmental education is considered a public policy that helps with socio-environmental issues and interacts with today's society, as Sorrentino and Trajber (2007) point out. Lipai et *al. (2007)* describe that according to Law No. 9/99 environmental education is conceptualised as a methodology in which the individual and the collective develop social principles, knowledge, attitudes and skills focused on environmental conservation, with human beings being responsible for the sustainability of the environment through their attitudes, whether public or private, individual or collective, is independent.

Figure 6 shows the presence of rubbish on the outskirts and inside the environmental preservation area, as well as children collecting recyclable materials that are contained in solid waste, the poor condition of the material used to fence off the area, with sabiá wood and wire, as well as signs of soil degradation, with the start of significant granulometric losses.

Cogo et al. (2003) state that one of the ways in which soil is most affected is by erosion, which reduces its productive potential and the environment, such as the silting up of water bodies, reducing the volumetric capacity for storing water, and water pollution, which jeopardises its use for a variety of purposes. Therefore, proper management of the water, soil and plant system through planning and conservation practices can control and reduce erosion.

Figure 7 shows the presence of animals in the preservation area in 2010, the toppling of fences, more intense erosion near the fencing of the area, which is already interfering with the community's area, and open trails within the forest that compromise its preservation and resilience.

Source: Author

Figure 7. Photographic record of environmental degradation in an environmental preservation area in Lagoa Seca/PB, 2010.

The photographic records show the degradation of the soil, with high levels of erosion such as gullies, the presence of people in the area and rubbish around the site.

The presence of a track shows the constant passage of people and animals, which directly damage the environmental area, such as burning, felling trees, grazing, soil compaction, among others (FIGURE 7).

The causes of soil degradation are diverse and water erosion is one of them. Vegetation cover also potentially interferes with soil erosion, so in a soil characterised as red-yellow latosol, water erosion losses were 0.26 t/ha over a six-month study period in a forest (SILVA et al., 2010).

The deforestation of species native to the environmental area is visualised in various stretches. It is constantly being aggressively interfered with by members of the Vila Florestal and Ipuarana communities, who remove trees for the use of wood, firewood and other utilities. Solid and liquid waste is constantly being dumped in the area. As well as being considered a Permanent Preservation Area due to its environmental attributes, at various points the area has a 45° slope, which also characterises it as a Permanent Preservation Area.

Environmental protection areas belong to the National System of Conservation Units, regulated by Law 9.985 of 18 July 2000. They are generally extensive areas, with a certain degree of human occupation, endowed with abiotic, biotic, aesthetic or cultural attributes that are especially important for the quality of life and well-being of human populations. Their basic objectives are to protect biological diversity, discipline the occupation process and ensure the sustainable use of natural resources, as well as being a source of seeds and presenting typical flora that is unique to each region.

Permanent Protection Areas are protected by federal law (BRASIL, 1965), but riparian forests in many places are degraded to a very aggressive level, especially in urban areas, as Martins (2001) states, and this damage to the environment leads to socio-economic loss and loss of quality of water resources, which are essential for supplying cities and sources of energy, according to Salvador (1986). As a result, given the length of time the law has existed, there should be more preserved areas and environmental awareness of public policies and environmental awareness of society as a whole.

Countries that are considered developed have various environmental plans in place, preserving natural resources and to the detriment of this action and encouraging the population to preserve and conserve nature, the benefits are varied. Basic water treatment is generally applied only to filtering, without the need to add any product, so the quality of the water is always maintained at levels that are favourable to the population, bearing in mind that water is a source of life and a vehicle for disease.

In order to better ensure conservation and the development of research, the importance of these "study units" must be included in local schools, as well as a great deal of mobilisation in the communities in relation to environmental education. In this line of thought, it can be seen that concern for the environment begins with education, which starts at home, so the school encompasses this knowledge, highlighting the importance of nature in the lives of human beings yesterday, today and tomorrow.

During the field visits, it was visually observed that the weir near the Preservation Area has an excess of sediment due to the lack of vegetation in the riparian forest, small reservoirs within the area are in the process of eutrophication, compromising the water demand for possible irrigation systems and the development of a seedling nursery.

Eutrophication can be defined as the accumulation of nutrients in water bodies, as stated by Figueirêdo et al. (2007) when this happens there is an exaggerated growth of aquatic plants, mainly cyanobacteria, with a consequent imbalance in the ecosystem and loss of water quality. According to some authors (CARMICHAEL, 1992; CARMICHAEL and FALCONER, 1993) the excess of cyanobacteria in water reservoirs causes various problems resulting from blooms that produce toxins and increase the occurrence of toxicity in people and animals, as well as making the site unsuitable for fishing.

Freitas et al. (2011) studied the loads of total phosphorus and suspended material in semi-arid reservoirs and found that these water bodies in semi-arid regions are seriously vulnerable to some negative impacts such as siltation, which results from the disaggregation of particles and is transported to the water sources through surface runoff, as well as eutrophication, one of the influential factors that degrades these water bodies due to the accumulation of nutrients.

Since water quality can be compromised due to a number of factors mentioned above, by collecting water samples from water bodies in and around the environmental preservation area, the results will expose the problems that occur in the region by checking the quality of the water, especially for irrigation purposes, which are used abundantly. Table 1 shows in detail the results of the water analysis, using physical and chemical parameters to give an overview of the environmental impacts generated.

Table 1. Water analysis of water bodies belonging to the environmental preservation area and its surroundings.

Water Analysis		
Parameters	**Weir**	**Spring**
PH	6,29	6,00
Conductivity (μS. Cm)-1	482	262
Calcium (meq/L)	0,87	0,52
Magnesium (meq/L)	1,34	0,28
Sodium (meq/L)	2,56	1,64
Potassium (meq/L)	0,38	0,18
Bicarbonates (meq/L)	1,86	0,74
Chlorides (meq/L)	2,9	1,92
Sodium Adsorption Ratio (SAR)	2,44	2,60
Water class	C2	C2

The pH of the water samples collected ranged from 6 to 6.29, Table 1, Ayres and Westcot (1991) state that the normal pH range for irrigation water is 6.5 to 8.4, i.e. the sample falls within the normal range. According to Sawyer et al. (1994), hydrogen potential (pH) is used to check whether a solution is acidic or basic through the concentration of hydrogen ions. Fluctuations in this parameter in waters considered to be natural can occur due to the consumption or production of carbon dioxide produced by photosynthesising bodies and respiration events in water bodies, which produce weak organic acids, according to Branco (1986).

When the pH of the water is basic, it can cause the precipitation of calcium and magnesium carbonates, according to Nakayama and Bucks (1986), which occurs at pHs above 7. Coelho et al. (2002) also states that the precipitation of fertilisers with low solubility can occur. pH values judged to be low do not wear out the irrigation system (MOURA et al., 2011). What can cause the pH to change are actions provoked by man, through the discharge of effluents of domestic and industrial origin (CETESB, 2005) and in the decomposition of pollutants in water bodies, pH variation can occur negatively (SPERLING, 1996).

The semi-arid region is characterised by irregular rainfall and high evaporation rates, which contribute to the concentration of salts in the soil and in the bodies of water that are the main source of irrigation water, mainly from rivers, dams and wells, so high salinity in irrigation water causes various problems with infiltration and plants, according to Andrade et al. (2007).

According to Silveira (2007), the pH of water used for irrigated agriculture is considered adequate when it ranges from 6.5 to 8.4. The water samples collected did not meet this standard, so when the water does not meet these values, it can lead to the degradation of irrigation equipment, as Ayres and Westcot (1999) state.

Ayers and Westcot (1999) highlight some problems related to the salinity of water for use in irrigation, such as the excess of soluble salts in the soil potentially reducing the availability of water for crops, due to soil sealing and toxicity, as cited by some authors (KUMAR et al., 2007; NAGARAJAN et al., 2010; REDDY and KUMAR, 2010).

The water analysed falls into class C2, i.e. water of medium salinity (EC between 250 and 750 $\mu S.Cm^{-1}$ at 25°C), as shown in Table 1. It can be used whenever there is a considerable degree of leaching and plants with moderate tolerance to salts can be grown in most cases without the need for special salinity control practices.

According to CETESB (2005) one of the characteristics that can be a factor that indicates the corrosiveness of water is high electrical conductivity values, as Fravet and Cruz (2007) announce that when electrical conductivity values exceed 100 $\mu S.cm^{-1}$ they are negatively reflected in the environment.

Lopes (2011) analysed water samples from the Olaria stream, in Pindorama/SP, and found that the values of the electrical conductivity parameter ranged from 80 to

120 μS.cm^{-1} , while in the São Francisco River the EC showed values of 40 to 92 μS.cm^{-1} (OLIVEIRA and HORN, 2006).

In studies carried out by Silva et al. (2012) to assess the water quality of rivers and reservoirs in the Paraíba River Basin, they found that the electrical conductivity varied from 0.06 dSm^{-1} to 2.2 μSm^{-1} . According to Ferreira et al. (2012), the electrical conductivity of the water sources belonging to the Paraíba River in the municipality of Amparo varied from 0.08 dSm^{-1} to 2.68 dSm^{-1} , being considered of good quality for human consumption.

In research carried out on bodies of water in the municipality of Camalaú/PB, Silva et al. (2014) observed electrical conductivity values and water salinity below 1.5 dSm^{-1} , with the exception of a point water sample located in the northern region, which had an EC of 1.9 dSm^{-1} , the sources with the highest conductivity values are cisterns and water wells, probably due to the EC being associated with the constituent minerals of the matrix rock. These same authors found that most of the samples were collected from surface water sources, which are more likely to be contaminated, but the water considered unsuitable for human consumption, according to the standards, came from wells.

As for sodicity, for values below three there are no restrictions on use according to the Ayres and Westcot classification (1991). Sodium is an important water quality parameter that can interfere with soil permeability. The effect of sodium is its toxicity to plants: high concentrations of this ion, when absorbed by plants, accumulate in the leaves, causing burns and necrosis on their edges.

Average magnesium concentrations ranged from a minimum of 0.28 in the reservoir to a maximum of 1.34 mg.L^{-1} in the spring near the reservoir. These values are well below the recommended limit for irrigation water, which ranges from 0 to 60 mg.L^{-1} (AYRES AND WESTCOT, 1999).

As for potassium, Ayres & Westecot (1991) comment that potassium concentrations ranging from 0 to 78 mg.L^{-1} are normal in irrigation water, values that were not exceeded in the research.

Silva et al. (2013), monitoring water sources in the Paraíba River for various purposes in the municipality of Barra de São Miguel, obtained results of adequate quality for human and animal consumption, with restrictions in only one sample analysed with levels above those permitted by the Ministry of Health.

After six years, a new visit was made to the site to check on the levels of degradation, and in 2016, as shown in Figure 8, it can be seen that the reservoir has no riparian forest and that various crops are grown around it, such as corn, sunflowers, banana trees, grass, among others. These agricultural sites are also used for experiments carried out by the university's researchers, who report that they address sustainability and organic production, although the conventional method is used and the water used for irrigation comes from the reservoir itself.

The water reservoir analysed shows the presence of algae (FIGURE 8), configuring the occurrence of eutrophication of the spring, so it can be seen that even after 3 years no recovery plan has been carried out in the area, continuing with the same problem of degradation of the riparian forest and water quality. The reduction in rainfall in the region at this time influences the quantity and quality of the water, and there is also evidence of sewage being discharged from the nearby community, increasing the concentration of pollutants and salts and potentially reducing its use for animals and irrigation, causing salinisation of the soil and a reduction in crop production.

Eutrophication, according to Silva et al. (2011), can be defined as a mechanism developed in the presence of a large organic load from domestic, municipal or industrial effluent discharges, which transport large quantities of nutrients into the reservoirs.

The high load of treated and untreated effluents discharged along watercourses that run close to urban perimeters or through large areas of agricultural production directly affect irrigators downstream, contaminating food and potentially the soil (Lucas et al., 2014).

The presence of animals in the surroundings was visually detected: cows and horses grazing in the area, trampling the soil and causing negative environmental impacts. On site visits, it was found that there is an excess of sediment in the weir due to the absence of vegetation in the riparian forest, soil degradation such as erosion and silting up of the water body (FIGURE 8), as also stated by Pereira et al. (2012).

Source: author

Figure 8. Degradation of water bodies and environmental preservation areas, 2016.

The presence of animals in the surrounding area is frequent, as can be seen in Figure 8 E , cows and horses being let loose to graze on the site, trampling the soil, which can lead to compaction, reducing its use and its potential for infiltration, generating surface water runoff into the weir, silting it up and reducing its total volumetric capacity. On site visits, it was found that there is an excess of sediment in the weir due to the absence of vegetation in the riparian forest, as was seen in 2010, but degradation has become more widespread over the years, requiring proper management planning for the preservation area.

In 2010, environmental degradation was observed in the environmental preservation area close to the site studied. Fragments of deforestation of native vegetation were noticeable in several places, and people from the community were removing wood and dumping solid and liquid rubbish in the area. This was also observed after six years, with the presence of waste in the area's surroundings,

trails in the environmental area, leftover bags of seedlings without a proper final destination, among other degradations that reduce the life expectancy of living beings in the environment.

Silva et al. (2013) studied the spring belonging to the watershed of the Piabas stream in the state of Paraíba, which is an area of environmental preservation, and found that anthropic actions are notable in the area, as well as the presence of grazing animals.

From the current physical characteristics of the area studied, it can be seen that it is environmentally disturbed due to the anthropogenic actions carried out, reducing the water storage capacity for the next rains. It will be necessary to restore the area in order to minimise the negative impacts generated.

Table 2 shows that there was no significant variation in the values of the water parameters analysed in relation to the water collection points. The pH varied from 5.82 to 6.04 $dS.m^{-1}$, making the pH of the water less than ideal. In water intended for irrigating annual or perennial crops, the appropriate pH range is between 6.5 and 8.4. Values outside this range can cause deterioration of irrigation equipment. In the surface reservoir analysed, the values are below the limit allowed for irrigation. Pereira et al. (2012) analysed the same reservoir in 2010 and obtained similar results with a pH of 6.29.

pH values far from neutral can affect aquatic life (ESTEVES, 1998). Dalastra et al. (2014), reports that the pH of water can be affected by the composition of the soil through which precipitation runs off the surface or percolates into the profile. Another source of pH alteration comes from anthropogenic actions, through domestic and industrial waste (CETESB, 2005). When the pH is far from 7.0, the microbial fauna responsible for the biological decomposition of pollutants can be negatively affected (SPERLING, 1996).

Table 2. Physical and chemical analyses of the water collected from the Lagoa Seca reservoir for irrigation purposes.

PARAMETERS	Sample 1	Sample 2
pH	5,82	6,04
Electrical conductivity (dS.m)$^{-1}$	1,72	1,53
Calcium (meq.L)$^{-1}$	2,25	2,19
Magnesium (meq.L)$^{-1}$	3,05	3,15
Sodium (meq.L)$^{-1}$	8,81	8,63
Potassium (meq.L)$^{-1}$	0,63	0,67
Carbonates (meq.L)$^{-1}$	0,00	0,00
Bicarbonates (meq.L)$^{-1}$	0,63	0,69
Chlorides (meq.L)$^{-1}$	11,30	11,72
Sulphates	Absence	Absence
Adsorption ratio of sodium - RAS	5,4	5,29
Water class	C3	C3

According to Ayres and Westcot (1991), the water in the Lagoa Seca reservoir has a slight to moderate use restriction due to its electrical conductivity being between 1-2.07 dS/m. Sodium is considered to have no use restrictions (<20 meq/L). Magnesium concentrations ranged from 3.05 to 3.15 meq.L^{-1} , at the source near the reserve. These values are well below the recommended limit for irrigation water, which ranges from 0 to 60 mg.L^{-1} .For the Potassium parameter, the authors Ayres & Westcot (1991) report that potassium concentrations ranging from 0 to 78 mg.L- 1 are normal in irrigation water, values not exceeded in the research.

Analysing RAS together with Ce to check the impact it can have on soil infiltration, it was found that there are no restrictions on its use. However, with regard to chloride > 3 meq. L^{-1} , there is a slightly moderate use restriction, and the water from the reservoir is in this situation. Chloride is not retained by soil particles, it moves easily with the water and is absorbed by the roots and translocated to the leaves, accumulating. When its concentration exceeds the plant's tolerance limit, damage occurs, such as necrosis and burning of the leaves. Vegetables are insensitive to foliar absorption even in water with high chloride concentrations, as explained by Ayres and

Westcot (1991). Water containing bicarbonates < 1.5, according to the guidelines for interpreting the quality of water for irrigation, has no restrictions on its use.

The water samples collected fall into class C3, according to Richards (1954) as can be seen in Table 2, i.e. water with high salinity, with a salt content of 700 to 2,250 micromhos/cm, cannot be used in soils with poor drainage and even with adequate drainage, special practices may be needed to control salinity and it should only be applied to irrigate salt-tolerant plants.

According to the Richards (1954) diagram for classifying water for irrigation, the sample of water from the reservoir analysed is classified as C3S1, with a high risk of use. However, irrigated crops must be tolerant to salinity and practise water and soil management techniques, making it necessary to calculate the need for leaching used during irrigation to reduce the concentration of salts in the soil and at the root of the plants. If this water is used incorrectly, this can lead to problems for the plants with reduced production and for the soil for subsequent agricultural use.

Silva et al. (2011) cite a number of problems that can be caused to crops when irrigated with inferior quality water with inadequate management, such as: excessive vegetative growth, delayed crop maturation, leaf and fruit stains, among others. Ghunmi et al. (2009) add the economic aspects due to the clogging of pipes with the accumulation of solids and the growth of microorganisms.

In the second survey, which looked at socio-economic issues as well as environmental ones, an assessment of socio-economic indicators was carried out in the community of Uruçu. The data collected showed that on average each family has around 5 members; 5 rooms in each house; the most common household appliances (television, fridge, cooker, radio, stereo, DVD, among others); most families don't have a bathroom in the house.

In the old days, toilets were built outside the home, which is what we found in this research: in order to use the toilet, you had to leave the house to go to the bathroom, regardless of the time of day, so people were used to this reality.

The water consumed for human consumption and cooking comes from the cisterns and the water used for non-noble purposes, such as some domestic services (washing clothes, bathing, etc.) comes mainly from wells and reservoirs. Photographic records show that after the cisterns were installed, there was a conceptual improvement in the problem of water scarcity. In the event of illness, they usually seek hospital treatment in Gurinhém or in the capital, João Pessoa/PB.

Brilhante et al. (2016), carrying out research into the quality of water in public school drinking fountains in Coremas/PB, found that all the water samples were contaminated by total coliforms and Escherichia coli, in addition to all the parameters assessed being outside the required potability standards (PORT.MS N°2.914/2011). However, Silva et *al.* (2014) analysed some water bodies located in Congo/PB and found that the water sources assessed were suitable for human consumption, without any restrictions, and were within the standards required by the Ministry of Health, and that water analysis is a resource for ensuring the population's quality of life.

In research carried out on the water sources of the Paraíba River by some authors (SILVA et al., 2012; FERREIRA et al., 2012) only one of the samples was considered unsuitable for consumption, being outside the quality standards with total solids values above the value allowed by the Ministry of Health, however the other water sources analysed are in accordance with the potability standards.

With regard to the residents, 43.75 per cent of all interviewees have a family income of between 1 and 2 minimum wages. 42.86% have primary education and 57.14% have secondary education. Of those interviewed, 95 per cent participate in community activities indirectly;

In the survey of teachers, 52.25 per cent of respondents said they had an income of around 1,700 reais. In terms of educational level, 11.1% have a postgraduate degree and 88.9% have an undergraduate degree. Of those interviewed, 100 per cent participate directly in actions linked to community development, from planning to implementing the restoration project.

Despite all the existing limitations, the school's teachers, coordinators and community leaders are continuing with the project to restore the area. According to the interviews, the last time seedlings were planted was during this year's Environment Week. The activity of raising awareness about natural resources and environmental education has been growing and developing more and more, and many teachers can already feel the difference in the students in various daily actions, in addition to the perception of the residents who work in other activities (farmers, housewives, shopkeepers, etc.) and who always take part in the events when they are held, because despite having studied little, they know the importance of conserving these resources, according to the various reports made spontaneously at the time of the interview.

The interviewees reported that after the construction of the cisterns there was an improvement in water supply, however, basic sanitation problems such as a lack of toilets and treated water were reported;

After three years, an evaluation was carried out in Uruçu at the site where the seedlings had been planted, and through photographic images, i.e. by visual means, the development of the attempt to recover the area through the planting of seedlings at the site was detected.

Thus, when comparing the time the seedlings were planted and after three years (Figure 9), the development of the plant species was satisfactory, with adequate height, and there was a large presence of vegetation around the site, with significant environmental recovery for the region, realising that the community's environmental awareness had great benefits for nature and for the inhabitants of the region themselves.

Fonte: SANTOS, R. T., 2008.

Fonte: Autor

Figura 9. Monitoring the development of planted plant species.

Pereira et al. (2012) when studying the survival and growth rate of tree species in the recovery of degraded areas, found that of the species used, Astronium fraxinifolium was the one that stood out with the lowest average, despite being considered ideal for reforesting degraded areas, which may have been due to its adaptability to the region, but its survival rate exceeded 80%. In this way, they concluded that when evaluating the species 18 months after planting them in the area to be recovered, the authors affirm that in the initial stage these species provide various advantages for the study area, due to their leaf area, soil protection by reducing the impact of precipitation directly on the soil, thus not wearing down the soil particles, and greater water infiltration.

Soils with low humidity and fertility should use Mimosa caesalpineafolia, as Carvalho (2003) explains, since it is the species that thrives best in places with these characteristics and gives satisfactory results when used to recover degraded areas. Plant species with a developed root system and large numbers of fine roots are important characteristics, due to the high absorption of nutrients which consequently increases growth (GONÇALES et al., 1992).

The species of the caatinga biome have several characteristics, one of which is their resistance to water stress, i.e. due to the dry seasons there are mechanisms in the plants that facilitate their survival in these environments, however any removal of vegetation cover, especially those considered woody, increases environmental degradation by reducing the percentage of survival of the young plants used in the recovery, (LIMA et al., 2015) which is why it is important to follow up in the first few years of implementing the seedlings in the area to monitor survival and also to check for any difficulties that may arise in this method of recovering the environment.

Lima et al. (2015) states that planting seedlings is one of the most suitable options for speeding up the process of recovering degraded areas, since the environmental circumstances are quite aggressive for the survival of seedlings that arrive naturally, requiring human intervention for tree maintenance to be carried out so that the largest number of seedlings planted can withstand the environmental conditions to which they will be subjected and have a higher percentage of life.

According to the interviewees at the "Chorona" spring, more than 2,000 seedlings were planted in the three years since the first joint effort, but only 5 per cent of them survived, due to the following factors: it is a private area belonging to farmers in Uruçu; the area is not fenced off, due to a lack of financial resources; the area is used for cattle grazing; and long periods of drought have compromised the development of the seedlings.

Figure 10 shows the initial conditions of the riverbed, with very aggressive silting, reducing the water storage capacity and its quality. However, after carrying out activities and the continued presence of researchers in the region, it was observed that over the years there has been a significant improvement, with greater volume capacity, visualisation of the stones that make up the riverbed and the perception of a resurgence of aquatic life, a result of the conservation of the area.

Fonte: JÚNIOR, J. A. S., 2006. Fonte: autor

Figure 10. Environmental conditions of the riverbed before and after the on-site study.

It is known that the fenced area is the main and pioneering action for a spring restoration project to be viable, as it isolates part of the anthropic interventions and must be carried out within the standards of permanent protection areas, according to the Forest Code, Law No. 7,803 of 18 July 1989; Act 2.

When recovering the vegetation cover of APPs that have already been degraded, a distinction must be made in terms of the type of water outcrop, i.e. without or with initial water accumulation, because the soil is waterlogged or temporarily submerged when it rains, the depth of the profile and the fertility of the soil are some of the factors that must be taken into account, as they are selective for the species that will be able to develop (RODRIGUES E SHEPHERD, 2000 apud COMITÊ DAS BACIAS HIDROGRÁFICAS DOS RIOS PIRACICABA, CAPIVARI, E JUNDIAÍ, 2004).

Borges (2010), working with restoration in Uberlândia-MG, found significant changes in the area after isolating it. The original vegetation quickly began to sprout and appear with exuberance. The water level in the stream bed increased, and it was possible to see all the places where the water flows and the formation of a main bed (in the lowest part of the erosion), making it clear that the vereda has great potential. In these parts where it is possible to notice the emergence of water, a dense layer of mosses and green algae has formed, helping to form a protective layer of soil and also inducing the germination of seeds that are deposited there when carried by the wind or animals. In this way, the fenced area of the vereda was enough to re-establish the development of the flora, stop the process of erosion and compaction due to cattle trampling and possibly re-establish, even if only partially, the volume of water in the spring.

Springs can be defined as water that emerges from the ground after the process of infiltration and percolation of precipitation into the soil, reaching the saturated zone (GOMES and VALENTE, 2005). However, the term spring is not appropriate, as the word means that the existence of water begins in that place, so the water that emerges in the spring is the resurgence of water on the earth's surface as a result of the processes of the hydrological cycle, being the point where the water generally has the best quality and minerals because it emerges from the rocks, and its preservation is important.

The inadequate use of natural resources, such as some practices applied by farmers for a long time, for example, burning, removing trees, among others, drastically affects the environment, especially water bodies and their springs, and may even extinguish them (XAVIER and TEIXEIRA, 2007). The destruction of natural resources, as reported by Biella and Costa (2006), influences the degradation of springs, leading to their extinction, which directly affects the constant supply of rivers.

According to environmental legislation, which came into force under Federal

Law No. 12,651/2012, known as the New Forest Code, where Permanent Preservation Areas (APPs) are the areas around springs and perennial waterholes, with a minimum radius of 50 metres, this dimension is not found in the areas studied.

According to CONAMA (1988), Environmental Protection Areas, also known as APAs, are conservation units designed to protect and conserve environmental quality and existing natural systems, with the aim of improving the quality of life of the local population and protecting regional ecosystems. This means that there must be ecological-economic zoning, which is generally not the case here.

Figure 11 shows the presence of developing seedlings in a spring area with reduced recovery capacity due to the presence of animals in the environment.

Source: author

Figura 11. Growing seedlings on site.

According to Ferreira et al. (2009), various methodologies are used to recover degraded areas with the aim of changing the scenario and reconstituting the area. Planting seedlings of native species is generally the most widely used procedure, according to Meneghello and Mattei (2004), as it enables recolonisation at a uniform density.

The lack of isolation of the area and the long periods of drought were the main factors in the loss of the seedlings planted. The work of the university through extension projects raises the potential of the community through the work of teachers, leaders and students.

A visit to the research site revealed that some crops are grown, such as maize (Figure 12). In order to plant them, the local forest has to be cleared to manage the soil and clean the area, and sometimes the practice of burning is still applied more frequently because it is a cultural issue and has been taught to family members over the years. The idea of

Burning is the fastest way to clean up the site and renew the soil, because after using this practice there is satisfactory crop production, but this procedure harms the living beings in the soil, reducing and diminishing them to the point where the soil's productive potential becomes almost nil, which is why it is important to use appropriate methods in soil management, in harmony with the environment and living beings, so that there is a balance in nature.

Water sources that have been degraded within the reserve due to excessive vegetation in their beds should be cleaned and deepened, as this is fundamental for maintaining local ecosystems. It is also important to use these water sources to irrigate seedlings and to control fires in the reserve itself.

The visits made revealed that the areas used for cultivation did not belong to small farmers, but rather to large local landowners (Figure 12), through land leasing, a very old practice that is still in use.

Source: author

Figura 12. Presence of agriculture on the site, growing maize.

Fagundes and Gastal Junior (2008), carrying out an environmental diagnosis and delimitation of permanent preservation areas in a rural settlement, found degradation in the APP's, indicating that revitalising riparian forests with native forests along the entire course of rivers helps to prevent siltation.

Visually (Figure 13) the forest was found to be in early succession in a permanent preservation area (APP), so we can see that the area is recovering and must be continually preserved in order to keep the area and its biotic and abiotic beings preserved for as long as possible. For this to happen, greater action would be needed, such as present supervision to maintain the preservation of nature, so the active participation of public policies and the community is fundamental so that by interacting with each other they can together be allies in the fight against environmental degradation. This is why laws are created so that they can be applied, but there is a need for greater encouragement for the population, environmental awareness and the application of punishments when degrading the environment in any way, so that there will be greater efficiency in environmental control and preservation.

Source: author

Figura 13. Permanent preservation area of the forest.

Figure 14 shows a large area of deforested land, with extensive removal of native vegetation, on a steep slope, which should not have been modified for any reason, as it is an APP, so for the nearby community it is a means of exploitation and survival. In a nearby area, a football match was played on an improvised pitch as a place for children to play, despite the extensive environmental damage.

Source: author

Figura 14. Deforestation in APP.

Rangel and Guerra (2014), when researching degradation in an ecological reserve, found severe degradation such as erosion on the slopes, as well as stating that the trampling that forms the trails is degradation due to soil compaction, reducing the capacity for water infiltration, verifying that the area studied is very negatively impacted, requiring recovery plans for degraded areas, realising the importance of field work to diagnose degradation and its possible causes, highlighting the importance of carrying out environmental research.

In order to recover the area more effectively, the area should be fenced off to make it more difficult for people and animals to enter the area, which contributes to the degradation of the native vegetation. The areas identified as being eroded on the banks of the reserve should be restored in order to prevent soil loss, minimise the increase in sediment deposits in the reservoir and control the erosion process.

CHAPTER 5

CONCLUSION

In the preservation area, environmental education work is needed in the surrounding communities to help with the sustainable management of the environment;

Regardless of the location of the preservation areas (Lagoa Seca or Uruçu), after the first environmental diagnosis it was realised that over the years environmental degradation still directly interferes with recovery planning;

The Environmental Preservation Areas mentioned in article 14 of the SNUC law (National System of Conservation Units, law no. 9.985 of 18/07/00), the APA (Environmental Protection Area) are part of the group of sustainable use conservation units and are the responsibility of the environmental agencies as potential managers and inspectors, however, this inspection is rather flawed.

The area is heavily impacted environmentally by various factors that degrade the site, such as the discharge of sewage, silting up, the absence of riparian forest, indiscriminate crop irrigation, among others.

It is necessary to carry out an area recovery plan so that the area can be rehabilitated for the benefit of the population and local biodiversity.

The water samples analysed from the Lagoa Seca reservoir can be used for irrigating salinity-tolerant crops, as well as applying appropriate management techniques to the irrigation system and the crop.

It is recommended to use leaching in irrigation to remove excess salts from the soil, as well as soil conservation techniques such as no-till and mulching.

BIBLIOGRAPHICAL REFERENCES

AB'SABER,A.N. The geoecological support of Brazilian (riparian) forests. IN: RODRIGUES, R.R.;LEITÃO FILHO, H.F. Matas Ciliares: Conservação e Recuperação. Edusp/Fapesp, 2001, p.12-25.

ALLEN,R. How to Save the World. London, Kogan Pagwe, 1980.

ALMEIDA, O. A . Irrigation Water Quality. Cruz das Almas: Embrapa Cassava and Fruit Growing, 2010.

ALMEIDA, O.A. Water quality for irrigation. Brazilian Agricultural Research Corporation, 2010.

ANDRADE, S. O. de. Impact of sewage from Riacho do Bode on the Piancó River - Pombal, PB. 2008, 38 f. Monograph (Graduation in Agronomy) Federal University of Campina Grande, Pombal, 2008.

ANDRADE, T. S.; SANTOS, E. S.; SILVA, J. R. L.; MONTENEGRO, A. A.; MONTENEGRO, S. M. G. L. Analyses of recharge and electrical conductivity in alluvium in the semi-arid region of Pernambuco. In: XVII Brazilian Symposium on Water Resources, 2007, São Paulo. Proceedings... Porto Alegre: Brazilian Water Resources Association, 2007. CD-ROM.

ARAUJO FILHO, J. A. A. Manipulation of the woody vegetation of the Caatinga for pastoral purposes. In.:Symposium on the Caatinga and its Rational Exploitation (SCER), 1984. Feira de Santana.

ARAÚJO FILHO, J.A. de; HOLANDA JÚNIOR, E.V.; SILVA, N.L. da; SOUSA, F.B. de; FRANÇA, F.M. Sistema agrossilvipastoril Embrapa Caprinos. In:LIMA, G.F. da C.;HOLANDA JÚNIOR, E.V.; MACIEL, F.C.; BARROS, N.N.; AMORIM, M.V.; CONFESSOR JÚNIOR, A.A. (Org.). Criação familiar de caprinos e ovinos no Rio Grande do Norte: orientações para viabilização do negócio rural. Natal: Emater, 2006. p.193-210.

ARAÚJO, A. E. de. Social Construction of Risks and Environmental Degradation: Municipality of Sousa, a case study. Master's dissertation - Department of Agricultural Engineering, Federal University of Paraíba, Campina Grande, 2002.

ARAÚJO, E. L.; SAMPAIO, E. V. S. B.; RHODAL, M. J. N. Floristic and phytosociological composition of three caatinga areas. Brazilian Journal of Biology. Vol. 55 n° 4, p. 595-607, 1995.

ARAÚJO, M. S.; XAVIER, L. A.; Identification of the proposal for the project to Recover Degraded Areas at the Seedling Production Centre of the State University of Paraíba Campus II Lagoa Seca. 2010.

ASSIS,R.L. Globalisation, sustainable development and local action: the case of organic farming. Cadernos de Ciência & Tecnologia. Vol. 20, n. 1, 2003, p. 79-96.

Atlas Brazil, 2013. United Nations Development Programme.

AVANZI, J. C.; BORGES, L. A. C.; CARVALHO, R. Legal protection of soil and water resources in Brazil. Revista em Agronegócios e Meio Ambiente, v. 2, n. 2, p. 115-128, 2009.

AYERS, R. S.; WESTCOT, D. W. A qualidade da água na agricultura. 2. ed. Campina Grande: UFPB, 1999, 153p. Irrigation and Drainage Studies 29.

AYERS, R. S.; WESTCOT, D. W. Water Quality in Agriculture. 2.ed. Campina Grande: UFPB, FAO, Irrigation and Drainage, 1999.

AYERS, R.S; WESTCOT, D. W. Water quality in agriculture. "Water Quality for Agriculture. FAO. Translated by Gheyi. H. R. & Medeiros, J.F de, UFPB.Campina Grande- PB, 217p. 1991.

BARBOSA,G.S.O DESAFIO DO DESENVOLVIMENTO SUSTENTÁVEL. Visões Magazine. Vol.1, n.4, 2008.

BARROS, J. D. S.; SILVA, M. F. P.;Sustainable Agricultural Practices as Alternatives to the Hegemonic Model of Production. Society and Rural Development online. Vol.4, n.2; 2010.

BEILLA, C. A; COSTA, R. A. Analysis of the Environmental Quality of the Urban Springs of Caldas Novas -GO. In: VI National Symposium on Geomorphology, Caldas Novas, 2006.

BELTRÃO,B.A. et, al. Project to register groundwater supply sources in the state of

Paraíba. Diagnosis of the municipality of Gurinhém. 2005.

BORGES, L.A.C.; REZENDE, J.L.P.; PEREIRA, J.A.A.; COELHO JÚNIOR. M.L.; BARROS, D.A. (2011).Permanent preservation areas in Brazilian environmental legislation. Ciência Rural, 41, n. 7, p. 1202-1210.

BORGES, W. J. Experience of recovering a degraded area located on the Irara farm, in the municipality of Uberlândia (MG).UFCG. Uberlândia - MG . 2010 (Monograph).

BRAZIL FOREST CODE Law N4.771, OF 15 SEPTEMBER 1965.Diário Oficial da União, Brasília: 18 September 1965.

BRAZIL. Law No. 12.608, of 10 April 2012.

BRAZIL. Law No. 4.771 of 15 September 1965 .

BRAZIL. Ministry of Planning. State Reform Master Plan. Brasília: Ministry of Planning, November 1995.

BRAZIL. NBR 9898. ABNT 1987. Available at: http://licenciadorambiental.com.br/wp-content/uploads/2015/01/NBR-9.898-Coleta-de-Amostampa.pdf. Accessed on 23.06.2016.

BRAZIL. Política Nacional do MeioAmbiente . Disponívelem : http://www.planalto.gov.br/CCIVIL/LEIS/L6938.HTM

BRASIL. Resolução CONAMAn° 001de 1986. Disponívelem : http://www.mma.gov.br/port/conama/res/res86/res0186.html

BRAZIL. National System of Nature Conservation Units. Available at: http://www.planalto.gov.br/ccivil_03/LEIS/L9985.htm

BRILHANTE, S. C., SANTOS, E. A., MARTINS, W. S., LEITE, C. M. F., SILVA, T. P. C., SOUSA, R. A., & MARACAJÁ, P. B. (2016). Microbiological and physical-chemical analysis of water from drinking fountains used in public schools in the city of Coremas-PB. Semiarid Technical Newsletter. Vol 10, n. 1, pag 05-08.

CADIER, E. Hydrology of small basins in the semi-arid north-east: Hydrological transposition. Recife, SUDENE/DPG/PRN/HME. (Hidrologia, 31). 448p. 1994.

CALDART, R. S. (Org): Dicionário da Educação do Campo / Organised by, Roseli Salet CALDART , I. B. P.; ALENTEJANO, P.; FRIGOTTO, G. (2012.). -Rio de Janeiro, São Paulo: Escola Politécnica de Saúde Joaquim Venancio, Expressão popular,

CAPPER, J. L.; CADY, R. A.; BAUMAN, D.E., The Environmental Impact of Dairy Production: 1944 Compared With 2007. J Anim Sci 2009. v.87: p.2160 - 2167Annals...Brasília: EMBRAPA - DDT 1986.

CARMICHAEL, W. W., AND I. R. FALCONER. Disease related to freshwater bluegreen algal toxins, and control measures. In Algal toxins in seafood and drinking water, ed. by I. R. Falconer, 187-209. San Diego, CA: Academic Press. 1993

CARMICHAEL, W.W. 1992. Cyanobacteria secmetabolites-the cyanotoxins. Journal of Applied Bacter, 72: 445-459.

CARPANEZZI, A.A. Indirect Benefits of the Forest. In: Reforestation of Rural Properties for Productive and Environmental Purposes. EMBRAPA Florestas, 2000.

CARVALHO, C.M.; ALMEIDA FILHO, R. Use of LANDSAT-TM images to assess the extent of desertification in the Gilbues region, southern Piauí state. In: Simposio Brasileiro de Sensoriamento Remoto, SC. INPE, p.4365-4372, 2007.

CARVALHO, P.E. R. Brazilian tree species. Colombo: EMBRAPA FLORESTAS (Technological Information), 2003. 1040 p.

CARVALHO,O. Semi-Arid Northeast: New Delimitation. Regional Bulletin of the Ministry of National Integration. National Regional Development Policy Newsletter. 2006.

CATELANI, C.S.; BATISTA, G.T.(2007).Mapping of permanent preservation areas (APP) in the municipality of Santo Antônio do Pinhal, SP: a subsidy for environmental preservation. Revista Ambiente & Água -An Interdisciplinary Journal of Applied Science, v. 2, n. 1.

CAVALHEIRO, F. Green areas: concepts, objectives and guidelines for planning. In: Proceedings of the 1st Brazilian Congress on Urban Afforestation, 13 to 18 September, 1992, Vitória.1992.

CEBALLOS,B.S.O. Use of Microbiological Indicators in the Typology of Aquatic Ecosystems in the Semi-Arid Tropics. PhD Thesis - Department of Hydraulics and Sanitation. São Carlos School of Engineering. University of São Paulo, São Paulo. 192p. 1995.

CETESB. ENVIRONMENTAL SANITATION TECHNOLOGY COMPANY. Report on the quality of inland waters in the state of São Paulo 2004/CETESB. São Paulo: CETESB, 2005.

Forest Code. Available at: www.planalto.gov.br: Accessed on 20/06/2011.

COELHO, E. F.; OLIVEIRA, A. S.; BORGES, A. L. Basic aspects of fertigation. In:BORGES, A. L.; COELHO, E. F.; TRINDADE, A. V. Fertirrigation in tropical fruit trees. Cruz das Almas: Embrapa Cassava and Fruit Growing, 2002. chap. 1, p. 9-14

COGO, N.P.; LEVIEN, R.; SCHWARZ, R.A. Soil and water losses by water erosion influenced by tillage methods, slope classes and soil fertility levels. Revista Brasileira de Ciências do Solo, v.27, n.1, p.743-753, 2003.

ENVIRONMENTAL SANITATION TECHNOLOGY COMPANY. Report on the quality of inland waters in the state of São Paulo 2002/ CETESB. São Paulo: CETESB, 2003. 264p.

BRAZILIAN CONGRESS ON URBAN AFFORESTATION, 1, 1992. Vitória.

Proceedings. Vitória: Brazilian Society of Urban Afforestation, p.15-28.1992.

NATIONAL ENVIRONMENTAL COUNCIL - CONAMA. RESOLUTION NO. 429, OF 28 FEBRUARY 2011.

CONTI, J. B. Clima e meio ambiente. São Paulo: Atual, 1998.

CPRM. Geological Survey of Brazil. Diagnosis of the Municipality of Gurinhém. Registration of groundwater supply sources project.2005.

CRISPIM, D.L.; LEITE, R.P.; CHAVES, A.C.G.; FERREIRA, A.C.; MEDEIROS, A.C.; MARACAJA, P.B. Impactos ambientais no açude do bairro nova vida em Pombal -PB. Revista Brasileira de Gestão ambiental, v.7, n.1, p.102-129, 2013.

DETZEL, V.A. Urban afforestation: importance and economic evaluation. In: Brazilian Congress on Urban Afforestation, 1, 1992. Vitória. Anais...Vitória: Brazilian Society of Urban Afforestation, p.39-52.1992.

DIEGUES, A. C. Ethnoconservation of nature: alternative approaches. In: Ethnoconservation: new directions for nature protection in the tropics. DIEGUES, Antonio Carlos (Org.), São Paulo: Annablume/Hucitec, 2000, p. 1-46.

DULLEY, R.D. Notção de Natureza, Ambiente, Meio Ambiente, Recursos Ambientais e Recursos Naturais gric. São Paulo, São Paulo, v. 51, n. 2, p. 15-26, jul./dez. 2004

DUMANSKI, J. A Framework for Evaluation of Sustainable Land Management (FESLM).Available: TDG website(1994). URL :http:// tdg.uoguelph.ca/www/FSR/collection/indicator/sus_land_manage.txt. Consulted on 12 January 2017.

EMBRAPA FLORESTAS (Colombo, PR). 2nd Embrapa Master Plan. Colombo, 2000.

EMBRAPA SEMI-ARID. 2nd Embrapa Semi-Arid Master Plan 2000-2003. Petrolina, PE, 2000. 55p.

ESTEVES, F. The difference between APA and APP is not clear to everyone. Chico Mendes Institute. 2011.

FAGUNDES, N.A.; GASTAL JUNIOR, C.V.S. Environmental diagnosis and delimitation of Permanent Preservation Areas in a rural settlement. Acta SCi. Biol. Sci, v.30, n.1, p.29-38, 2008.

FARIAS, M. S.; LIMA, V. L. A. Water Resources. In: ROCHA et al. Manejo Ecológico Integrado de bacias hidrográficas no semiárido brasileiro. Campina Grande: Epgraf, 2011. 332 p.

FERREIRA, A. C.; SILVA, V. F.; MEDEIROS, J. X. ; BARACUHY, J. G. V.; LIMA, V. L. A. Study of water quality in the municipality of Amparo-PB. 7th Congress of Higher Agricultural Education and 52ª Annual Meeting of Abeas 29 October to 01 November 2012.

FERREIRA, R.A. et al. Direct seeding with forest species in the implantation of riparian forest in the Lower São Francisco in Sergipe. Sciencia Forestalis, Piracicaba, v.37, p.37-46, 2009.

FIGUEIREDO, M.C.B.; TEIXEIRA, A.S.; ARAÚJO, L.F.P.; ROSA, M.F.; PAULINO, W.D.; MOTA, S.; ARAUJO, J.C. Evaluation of the environmental vulnerability of reservoirs to eutrophication. Engenharia sanitária ambiental, v.12, n.4, p.399-409, 2007.

FIRMINO, R. G.; FONSECA, M.B. da. An Economic View of the Environmental Impacts Caused by the Expansion of Agriculture. In: Extension Meeting 10. UFPB-PRAC, 2008.

FRAVET, A. M. M. F.; CRUZ, R. L. Quality of water used for irrigating vegetables in the Botucatu-SP region. Revista Irriga, Botucatu-SP, v. 12, n. 2, p. 144-155, 2007.

FREDERICO, C.; FURTADO, D. A.; Integrated Water Resources Management, p.27. Hydrometry Course - UFCG.

FREITAS, F.R.S.; RIGHETTO, A.M.; ATTAYDE, J.L. Total phosphorus loads and suspended matter in a Brazilian semiarid reservoir. Oesologia Australis, v.15, n.3, p.655-665, 2011.

GARJULLI, R. Water resources in the semi-arid region. Ciência e Cultura. v.55, n.4, 2003.

GLIESSMAN, S. R. Agroecology: ecological processes in sustainable agriculture. Porto Alegre: Ed. Universidade/UFRGS, 2000.

GOMES, M.A.; VALENTE, O.F. Conservation of springs: Hydrology and Management of Headwater Basins, Viçosa, MG: Editora Aprenda Fácil, 2005. 210 p.

GONÇALVES, J. L. M. et al. Biomass production and root system of species of different successional stages. Revista do Instituto Florestal, v.4, p 363-367, 1992

GOODLAND, R. & LEDOC, G. "Neoclassical Economics and Principies of Sustainable Development". Ecological Modelling, 38, 1987.

GOVERNMENT OF THE STATE OF PARAIBA. Geographical Atlas of the State of Paraíba. Paraíba State Secretariat / UFPB. João Pessoa, PB. 100p. 1985.

GUTIÉRREZ, J.R. & SQUEO, F.A. Importance of shrubs in semi-arid ecosystems in Chile. Ecosistemas, 2004/1

HAGMAN, G. Prevention Better than Cure, Report on Human and Environmental Disasters in the Third World, Prepared for the Swedish Red Cross, Stockholm. 1984.

INSTITUTES THE NATIONAL CIVIL PROTECTION AND DEFENCE POLICY. A vailable at <http://www.planalto.gov.br/ccivil_03/_Ato20112014/2012/Lei/L12608.htm>. Accessed on 09/03/2016

JACOB, P. Cadernos de Pesquisa, n. 118. March 2003

K. MARZALL & J. ALMEIDA. Sustainability indicators for agroecosystems. Caderno de Ciências e Tecnologia, v.17, n.1, p.41-59, 2000.

KAZIOSKI, G.V.; CIOCCA, M. L.S. Energy and Sustainability in Agroecosystems. Ciência Rural, v. 30, n.4, 2000, p.737 -745.

KUMAR, M.; KUMARI, K.; RAMANATHAN, A. L.; SAXENA, R. A comparative evaluation of groundwater suitability forirrigation and drinking purposes in two intensively cultivated districts of Punjab, India. Environmental Geology, Berlin, v.53, p.553-574, 2007.

LEMOS, J.J.S. Levels of Degradation in the Brazilian Northeast. Revista Econômica do Nordeste, v.32, n.3, p.406-429, 2001.

LIMA, J. R. de L.a et al., (Coord.). National action programme to combat desertification and mitigate the effects of drought - PAN BRASIL. Brasília - DF: Ministry of the Environment / Secretariat of Water Resources, 2004.

LIMA, M. M., DE ARAÚJO SANTOS, L., MOURA, F. D. B. P., & DE SOUZA NOGUEIRA, E. M. (2015). Initial survival of six species used in the recovery of a degraded area in the caatinga. Revista Ouricuri, 5(2), 132-137.

LIMA, V. de P. Hydrological function of riparian forest. In: SYMPOSIUM ON CYLINARY FOREST, 1., 1989, São Paulo. Proceedings... Campinas: Cargill Foundation, 1989. p. 2542.

LIMA, V. L. A.; Training Course in Hydrometry of the Micro-Hydrographic Basins of the Brazilian Semi-Arid. Water Quality Module. 2010. 57p.

LIMA, W. de P.; ZAKIA, M. J. B. Hydrology of Riparian Forests. In: Riparian Forests Conservation and Recovery. São Paulo. Editora Universidade de São Paulo: Fapesp, 2000.

Link:http://pt.wikipedia.org/wiki/%C3%81rea_de_prote%C3%A7%C3%A3o_ambiental. Accessed in December 2011.

LIPAI, E.M.; LAYRARGUES, P.P. PEDRO, V.V. Educação Ambiental na escola: ta na lei. Let's take care of Brazil: concepts and practices in environmental education at school. Ministry of Education, General Coordination of Environmental Education: Ministry of the Environment, Department of Environmental Education: UNESCO, 2007.

LIRA, E. H. A. de.Floristic survey and titosociological analysis of a remnant of Atlantic forest located in the municipality of Lagoa Seca-PB. Lagoa Seca -PB/ Emannuella Hayanna Alves de Lira. -Course Conclusion Work (Graduation in Agroecology) - Paraíba State University. Centre for Agricultural and Environmental Sciences, 2013.

LIRA, W.S.; CANDIDO, G.A. Analysis of Indicator Models in the context of sustainable development. Perspectivas Contemporâneas, v.3,n.1,p.31-45, 2008.

LOPES, A.S. & GUILHERME, L.R.G. Efficient use of fertilisers and agricultural correctives: agronomic aspects - A.S. Lopes and L.R.G. Guilherme - 3rd revised and updated edition - São Paulo, ANDA, 2000

LOPES, C. L. Monitoring water quality in the Olaria stream watershed. Fórum Ambiental da Alta Paulista, São Paulo-SP, v. 7, n. 2, p. 452-461, 2011.

LOUZADA, F. L. R. O.; VIEIRA, M. V. M.; PELUZIO, T. M. O.; SAITO, N. S.; SOUZA, S. M.; SANTOS, A. R.(2009).Utilisation of GIS in the determination of APP on hilltops in the planning micro-region of the Linhares pole, ES. In: Encontro Latino Americano de Iniciação Científica, Encontro Latino Americano de Pós-Graduação, Encontro Latino Americano de Iniciação Científica Júnior, 12; 9; 3, São José dos Campos -SP. Proceedings...São José dos Campos -SP: UNIVAP, 2009.

LUCENA, J.H. de. Study of the Spatial-Temporal Variation of Water Quality in a stretch of the Bodocongó River, PB. Master's dissertation. 1998.

MACHADO, C, J. S. et al. Environmental legislation and environmental degradation of soil by oil activity in Brazil. Desenvolvimento e Meio Ambiente, v. 28, p. 41-55, Jul./Dec. 2013

MANTOVANI, W. Conceptualisation and conditioning factors. In: SYMPOSIUM ON CYLINARY FOREST. Proceedings. Campinas: Cargill Foundation, 1989. p 11-19

MARTINS, S.V. Recovery of Riparian Forests. Viçosa: Aprenda Fácil, 2001. 143p.

MELO FILHO, J. F. de; SOUZA, A. L. V.. Soil Management and Conservation in the Semi-Arid Region of Bahia: Challenges for Sustainability. Bahia Agrícola, v. 7, n. 3, pg. 50-60, 2006.

MENDES, B. V. Faunal Resources. Brazilian Association for Higher Agricultural Education - ABEAS. Brasília: DF. ABEAS. Campina Grande: UFCG, Sustainable Development Course for the Brazilian Semi-Arid. Module 10. 2006.

MENEGHELLO, G. E.; MATTEI, V. L. Direct sowing of Timbaúva (Enterolobium contortisiliquum), Canafístula (Peltophorum dubium) and Cedar (Cedrela fissilis) in disturbed fields. Ciências Florestal, Santa Maria, v.14, n.2, p. 21-27. 2004.

MENEGUZZO, I. S. Analysis of environmental degradation in the urban area of the Arroio Gertrudes basin, Ponta Grossa, PR: a contribution to environmental planning. Master's dissertation. Postgraduate Programme in Soil Science, Agricultural Sciences Sector, Federal University of Paraná, 2006, 99 f.

MENEGUZZO, I. S.; Chaicouski, A. Reflections on the concepts... Geografia (Londrina) v.19, n.1, 2010. http://www.uel.br/revistas/uel/index.php/geografia/ 185

MILANO, M.S. The city, open spaces and vegetation. In: Encontro Nacional Sobre Arborização, 4, Vitória, 1992. Anais...Vitória, Vitória City Hall, p. 317-326.1992.

MINISTRY OF THE ENVIRONMENT (2011). Areas affected by the tragedy of the rains in the mountainous region of Rio de Janeiro: permanent preservation areas and conservation units x risk areas - what does one have to do with the other? Brasília:

MMA.

MINISTRY OF THE ENVIRONMENT. 2016. http://www.mma.gov.br/areas-protected/units-of-conservation/what-they-are

MIRANDA, J.R. Ecologie des peuplements de repiles du Tropique SemiAride Bresilien (région d'Ouricuri-PE). Montpellier: Academie de Montpellier- Université des Sciences et Techniques du Languedoc. Ecole Pratique des Hautes Etudes, 1986. 418p. Doctoral dissertation (Speciality: Biology of Populations and Ecosystems).

MONTEIRO, R. S. (2002). Environmental education in Mato Grosso / Roselane Soares Monteiro -Brasília: Ministry of National Integration: Federal University of Mato Grosso

MORAES, D.S.; JORDÃO, B.Q. Environmental degradation and human health. Revista saúde publica, v.36, n.3, p.370-374, 2002.

MOURA, R.S.; HERNANDEZ, F. B. T.; LEITE, M. A.; FRANCO, R. A. M.; FEITOSA,D. G.; MACHADO, L. F. Water quality for irrigation use in the Córrego do Cinturão Verde watershed, municipality of Ilha Solteira. Revista Brasileira de Agricultura Irrigada, Fortaleza-CE, v. 5, n. 1, p. 68-74, 2011.

MUNIZ, F. H. 2004. The vegetation of the transition region between Amazonia and the Northeast, diversity and structure. In: MOURA, E. G., coord. Agroenvironments of the transition between the tropics and the semi-arid region of Brazil. State University of Maranhão, São Luis, Maranhão. p.53-69

NAGARAJAN, R.; RAJMOHAN, N.; MAHENDRAN, U.; SENTHAMILKUMAR, S. Evaluation of groundwater quality and its suitability for drinking and agricultural use in Thanjavur city, Tamil Nadu, India. Environmental Monitoring and Assessment, The Netherlands,v.171, p.289-308,2010.

NAKAYAMA, F. S.; BUCKS, D. A. Trickle irrigation for crop production. Amsterdam: ELSEVIER, 1986. 383p

NASCIMENTO, E.P. Trajetória da sustentabilidade:do ambiental ao social, do social ao econômico estudos avançados,v.26, n.74,2012.

OLIVEIRA, E. M.; SOUTO, J. S. Edaphic mesofauna as an indicator of degraded areas. Revista Verde de Agroecologia e Desenvolvimento Sustentável, v. 6, n. 1, p. 1-9, 2011.

OLIVEIRA, M.R. de.; HORN, A.H. Comparison of the concentration of heavy metals in the waters of the São Francisco River in Três Marias, from 1991 to the present, in relation to the actions of the Cmm- Três Marias. Geonomos, Três Marias-MG, v.14, n. 2, p. 55-63, 2006.

OLIVEIRA, M.Z.; VERONEZ, M.R.; THUM, A.B.; REINHARDT, A.O.; BARETTA, L.; VALLES, T.H.A.; ZARDO, D.; SILVEIRA, L.K. (2007). Delimitation of permanent preservation areas: A case study using high-resolution satellite imagery associated with a geographic information system (GIS). In: Brazilian Symposium on Remote Sensing, 13, Florianópolis. Anais ..., Florianópolis: SBSR, 2007 p. 4119-4128.

PAIS, P.S.M.; SILVA, F.F.; FERREIRA, D.M. Environmental degradation in the State of Bahia: an application of multivariate analysis. Revista Geonordeste, v.23, n.1, p.1- 12, 2012.

PEREIRA, J.S., DE ABREU, C.F.N.R., JUNIOR, R.A.P., & RODRIGUES, S.C. (2016). Evaluation of the survival and growth rate of tree species used in the recovery of degraded areas. Revista Geonorte, 3(4), 138-148.

PEREIRA, J.S.; SILVA, V.F.; PEREIRA, F.J.; FARIAS, M.S.S. Diagnosis of an environmental preservation area (APA) in the municipality of Lagoa Seca-PB for recovery purposes. p.430-433, 2012.

PEREIRA, J.S.; SILVA, V.F.; PEREIRA, F.J.; FARIAS, M.S.S. Diagnosis of an environmental preservation area (APA) in the municipality of Lagoa Seca-PB for recovery purposes. p.430-433, 2012.

PHILIPPI, A. JR; ROMERO, M.A.; BRUNA, G.C. Curso de gestão ambiental. Barueri, SP: Manole, 2004.

PINTO, N.G.M.; CORONEL, D.A. Environmental degradation in Rio Grande do Sul: an analysis of cities and mesoregions. Brazilian Journal of Agroecology, v.9, n.1, p.3-17, 2014.

QUEIRÓZ, M. A. de. Potential of Cucurbitaceae germplasm in the Brazilian Northeast. Horticultura Brasileira. v. 11, n. 1, p. 7-9, 1993.

RANGEL, L.A.; GUERRA, A.J.T. Degradation of trails in the Juatinga Ecological Reserve in Paraty-Rio de Janeiro. Rev. Ambient. Água,vol. 9,n. 4,. 2014.

REDDY, A.G.S.; KUMAR, N.K. Identification of the hydrogeochemical processes in groundwater using major ion chemistry: a case study of Penna-Chitravathi river basins in Southern India. Environmental Monitoring and Assessment, The Netherlands, v.170, p.365-382, 2010.

RESOLUTION/CONAMA/No. 010 of 14 December 1988

RICHARDS. L.A. (Ed.). Diagnosis and rehabilitation of saline and sodic soils. 5. ed. México: Centro Regional de Ayuda Técnica, 1970. 172p. il (Centro Regional de Ayuda Técnica. Manual de Agricultor, 60).

ROCHA, J. S. M. da. Vegetation Diagnosis. In: CIPAM Project. UFSM, 2005.

ROSSATO, M.V. Qualidade ambiental e qualidade de vida nos municípios do Estado do Rio Grande do Sul. 2006. Thesis (doctorate in applied economics) - Federal University of Viçosa-MG.

SAADI, A. The backlands that became deserts. B. Inf. SBCS, 25:1:10-17, 2000.

SALCEDO, I. H. Soil fertility in subsistence agriculture: Challenges for the semi-arid Northeast. In: BRAZILIAN MEETING ON SOIL FERTILITY AND PLANT NUTRITION, 26. BRAZILIAN MEETING ON MYCORRHIZAE, 10.

BRAZILIAN SYMPOSIUM ON SOIL MICROBIOLOGY, 8. BRAZILIAN MEETING ON SOIL BIOLOGY, 5. 2004, Lages. Proceedings... Lages, 2004 CDROM.

SALES, M. C. L. Evolution of desertification studies in northeast Brazil. GEOUSP - Espaço e Tempo, n. 11, p.115-126, 2002.

SALVADOR, J.L.G. Behaviour of native forest species in reservoir depletion areas. IPEF, Piracicaba, n.33, p.73-78, 1986.

SAMPAIO, E.V.S.B.; ARAUJO, M.S.B.; SAMPAIO, Y.S.B. Impactos ambientais da agricultura no processo de desertificação no Nordeste do Brasil. Revista de Geografia, v.22, n.1, p.93-113, 2005.

SÁNCHEZ, L. E. Environmental Impact Assessment: concepts and methods. São Paulo: Oficina de Textos, 2008.

SANTANA, M. J.; CARVALHO, J. A.; SOUZA, K. J.; SOUSA, A. M. G.; VASCONCELOS, C. L.; ANDRADE, L. A. B. Effects of irrigation water salinity on the sprouting and initial development of sugarcane (Saccharum spp) and on soils with different textural levels. Revista Ciência Agrotécnica, v.31, p.1470-1476, 2007.

SANTOS, J.A.J.; BARROS, G.J.; SANTOS, R. T; LIMA, J. K. S.; BRITO, E. T. F. S. Resíduos Sólidos: o desafio da Educação Ambiental na comunidade rural de Uruçú - Gurinhém/PB. Revista Educação Ambiental em Ação, Gurinhém/PB, 2011. v.5, 2011.

SANTOS, M. E. P. dos. Some considerations on the concept of sustainability: its political, theoretical and ontological dimensions. In: RODRIGUES, A. M. Desenvolvimento sustentável, teorias, debates e aplicabilidades . Campinas: UNICAMP/IFCH, 1996. p. 13-48. (Textos Didáticos, n. 23).

SANTOS, R. T.; Individual Report of the Citizen Universities Project - UFCG, June 2008.

SATTLER, M. A. Urban afforestation and environmental comfort. In: CONGRESSO BRASILEIRO SOBRE ARBORIZAÇÃO URBANA, 1; ENCONTRO NACIONAL SOBRE ARBORIZAÇÃO URBANA, 4, Vitória, 1992. Proceedings. Vitória: City Hall, 1992. V.1, p. 15-28.

SAUER, C.A. Permanent preservation areas: legislation, preservation and environmental restoration in riparian areas. Monograph (Specialisation in Environmental Law)-Universidade regional do Noroeste do Estado do Rio Grande do Sul, Ijuí, 2008.

SILVA, H. L. Interpretação qualitativa da evasão no contexto escolar: o caso do Centro de Estudos Supletivos de Vitória. Vitória, 1987. Dissertation (Master's Degree) - UFES.

SILVA, J.V.; ALECRIM, M.A.B.; SILVA, D.O.; COSTA, C.C.; OLIVEIRA, R.J. Soil and water losses due to water erosion in an equiaxial forest on a red-yellow latosol. Revista Brasileira de Ciencias Agrarias, v.5, n.5, 2010.

SILVA, L.F.T.C.; CASTRO, U.N.; GUERRA, A.J.T.; LIMA, F.S.; MENDES, S.P.; BEZERRA, J.F.R. Environmental degradation in areas destined for cattle ranching in the Sana River sub-basin, Macaé-RJ. Revista de Geografia, v.29, n.1, p.45-59, 2012.

SILVA, M. A. V.; BRAGA, C. C.; NIETZSCHE, M. H. Atlas climatológico do Estado da Paraíba. 2. ed. Campina Grande: UFPB, 1987.

SILVA, O. J. da; Rainwater Harvesting in the City of Campina Grande, PB. Alternative for a Policy to Deal with Water Shortages in Public Schools. Master's dissertation, 2003.

SILVA, R. C. S.; ALMEIDA, J. C. R.; BATISTA, G. T.; FORTES NETO, P. Physical, chemical and biological indicators of soil quality and the sustainability of natural environments. Electronic Repository Agricultural Sciences, Environmental Sciences Collection, p. 1-13, 2011

SILVA, V. F., ALBA, L. D. A., FERREIRA, A. C., BARACUHY, J. G. D. V., & MARACAJÁ, P. B. (2014). Evaluation of the quality of the water bodies of the paraíba river in cariri paraibano. Green Journal of Agroecology and Sustainable Development, 8(3), 235-238.

SILVA, V. F., ALBA, L. D. A., FERREIRA, A. C., BARACUHY, J. G. D. V., & MARACAJÁ, P. B. (2014). Potability of water in hydric sources belonging to the paraíba river. Green Journal of Agroecology and Sustainable Development, 9(1), 238-243.

SILVA, V. F.; BRITO, K. S. A de.; PEREIRA, J. S.; FERREIRA, A. C.; BARACUHY, J. G. V. Monitoring of water sources in the Paraiba River. IN:IV International Meeting on Water Governance: Innovation in Water Governance and Climate Variations in the Ibero-American Context. São Paulo-SP, 2013. ISBN: 978-85-86923-31-9.

SILVA, V. F.; FERREIRA, D. J. L.; FERREIRA, A. C.; MEDEIROS, J. X. ; BARACUHY, J. G. V.; LIMA, V. L. A. Monitoring the water quality of water bodies in the upper Paraíba region. In: Brazilian Symposium on Rainwater Harvesting and Management. Campina Grande- PB, 2012.

SILVA, V.F.; FERREIRA, A.C.; LIMA, V.L.A.; BARACUHY, J.G.V. Diagnosis of the spring in the Piabas Creek watershed. 1st International Workshop on Water in the Brazilian Semi-Arid. Campina Grande, 2013.

SILVEIRA,T. Physical-chemical analysis of water from the Cabe River Basin

SORRENTINO, M.; TRAJBER, R. Environmental Education Policies of the Governing Body. Let's take care of Brazil: concepts and practices in environmental education at school. Ministry of Education, General Coordination of Environmental Education: Ministry of the Environment, Department of Environmental Education : UNESCO, 2007

SOUSA, O. R. História Geral: da pré-história aos últimos factos dos nossos dias, 24 ed. São Paulo, Ática, p. 5-375, 1985

SUERTEGARY, D. M. A. Desertification: Recovery and Sustainable Development. In: CUNHA, S.B. & GUERRA, AJ.T. Geomorfologia e Meio Ambiente. RJ: Bertrand Brasil, 1996.

TAVARES, S.R.L. Recovery of degraded and/or contaminated areas: a challenge for the agricultural sciences. II symposium of agrarian sciences in the Amazon. Santarém- PA, 2015.

TUNDISI, J. G. Water in the 21st century: facing scarcity. São Carlos: RIMA, IIE, 248p. 2003.

VERDUM, R.; FONTOURA, L.F.M.Rural territories: from place to record. 2009.

VIANNA, L.S.; SILVA, K. G.; BERTOSSI, A. P. A.; MENDES, T. N.; XAVIER, T. M. T. ANALYSIS OF WATER QUALITY FOR IRRIGATION PURPOSES IN THE RIO ALEGRE WATERSHED, ES. ENCICLOPÉDIA BIOSFERA, Centro Científico Conhecer, vol.7, N.12; 2011.

VIEIRA, V. C. B. LIMA, M. G. Diagnóstico hidroclimático do município de Gilbués-PI, II CONNEPI, João Pessoa-PB, p. 1-10, 2007.

VON SPERLING, M. Introdução a qualidade das águas e ao tratamento de esgotos. Belo Horizonte: Department of Sanitary and Environmental Engineering/UFMG, 1996. 246p.

XAVIER, A. L.; TEIXEIRA, D. A. Diagnosis of the springs in the São João river sub-basin in Itaúna, MG. In: CONGRESS OF ECOLOGY OF BRAZIL, 7, 2007, Caxambu. Proceedings... Caxambu: SEB, 2007. p. 1-2

Printed by Books on Demand GmbH, Norderstedt / Germany